Your Healthy Brain

A Personal & Family Guide to Staying Healthy & Living Longer

Stephen J. Kiraly, MD, FRCPC

CONTENTS

PREFACE

Your Healthy Brain is a beginners book which will suffice unless you plan to go to medical school. The first edition of this book was a bit too technical and received the appropriate corrective criticism. The aim is to teach a limited range of knowledge around brain and brain functions in a simple and effective manner. To achieve this end, there have been some simplifications and restricted vocabulary with some efforts to bring definitions into present-day practice and knowledge of physiology. This limitation of subject matter will give the reader a better chance to assimilate basic knowledge and to make good use of it. Too often those interested in brain physiology are rushed through a great deal of scientific and pseudoscientific material of which they will retain very little. This spells ruin and loss of interest in the subject; they will never get a hold of it and find it always too difficult and complicated to grasp. In this book the material is limited to what the general public, with a little work, can assimilate and use.

My aim is to reflect accurate knowledge and to translate that into healthy practices. In all the chapters there is a large element of discrete review and repetition to make a point. There are some instructions to follow which are beneficial exercises and promote further exploration and learning. I try to stay practical and included a glossary to explain scientific and medical terms.

The book may be used by a facilitator to teach groups or to use in a classroom format. Interaction would certainly help maintain the interest of the participant.

Your Healthy Brain is the fruit of long experience and of careful planning and execution. It is my earnest hope that this book will prove informative and helpful to readers and teachers alike.

Stephen J. Kiraly, MD, FRCPC

To Barbara

Acknowledgments

Since the initial interview with the British Columbia chapter of Mensa (thanks to Stuart Munro), and the first workshop in Vancouver (Canada) in 2000, I, with the help of others, as acknowledged in the first edition, have done a huge amount of work in delivering hundreds of presentations on *Your Healthy Brain*. Finally, after much badgering from friends, colleagues and most of all patients, the *first edition* appeared in 2007 and was snapped up by local bookstores and workshop attendees. My clinical commitments in Geriatric and Adult Psychiatry have been too onerous to start on a second edition, but I am worn down again by popular demand to produce an update. *Viola – the second edition!*

Fundamental to the whole project were the seminar and workshop participants whose cumulative input and constructive criticisms were invaluable and provided me with ongoing critique, stimulation and education. Those people and groups were well recognized in the first edition; therefore, their names will not be reiterated here.

However, special mention is deserved by Dr. Stephen Holliday who helped with the writing of the first edition and Dr. Gulzar Cheema, former Minister Responsible for Mental Health, who delivered the opening remarks and welcome for the 2nd Annual Conference (2002) on the Healthy Brain Program for the family doctors of British Columbia.

I also owe special thanks to Dr. Pat McGeer, trail-blazing researcher in neurodegenerative disease at the University of British Columbia.

FOREWORD

It is a pleasure to recommend this unique book to readers interested in maintaining a healthy brain. Health professionals of all types, as well as fitness experts, have always taken the brain for granted, as though it needed no attention. Do everything possible to look after the lungs, the heart, the joints, the muscles, and other organs, but forget about the brain since it must be able to look after itself. Nothing could be further from the truth. The brain is the most important organ of the body and its fitness should never be taken for granted. It gives rise to the mind, so a healthy brain generates a healthy mind. Looking after the brain should be at the top of our healthy lifestyle list. But, all too frequently, it is at the bottom. The consequences are frequently tragic. The brain has no pain fibres, so the warning signals we respond to elsewhere in the body are absent. That is all the more reason why we should give our brain loving care.

What principles should we follow to care for our brain? Until this book, advice was almost non existent. But Stephen Kiraly, as a specialist in geriatrics and adult psychiatry, has spent much of his professional life analyzing what leads to an unhealthy brain, and therefore what we should do to maintain a healthy brain.

He has developed easy methods for us to follow. His Healthy Brain Program has been presented, largely to professionals, at many workshops over recent years. Now he has brought it out in a book that is easy to read and suitable for the lay public. Everyone can put a copy on their shelves, hopefully where it can be referred to frequently. As a physician and brain scientist I can only wonder why it has taken the medical profession so long to begin addressing such a vital subject. This book is a first and I heartily endorse it.

Dr. Patrick McGeer, OC, OBC, MD, PhD, FRSC
Professor Emeritus, UBC Faculty of Medicine, 2007
Director, Kinsmen Laboratory of Neurological Research

INTRODUCTION

When I first became interested in the brain health project, almost 15 years ago, I was caught up with so much enthusiasm that I came to believe what I was preaching. That was a crucial mutative event for without that belief, no change would have taken place. The result was that I lost 15 lbs of overweight, became more fit, quit smoking and put a higher value on managing stress, nutrition and recreation; in short – a better life. Now, I don't want to mislead you, the reader, as I am 15 years older and I can feel it. Brain program or no brain program, we will not live forever. The Healthy Brain Program© is a method, not a fad or merely a business idea; it has become a grass roots movement that is now supported by all sane leaders, politicians and citizens. What's more, increasing awareness of the degradation and even looming disasters awaiting our Earth dictate a brainy solution to brainless living.

I had the fortunate experience of delving through the laboratory and the intellectual productions of other people who have done significant research in many areas pertaining to the brain. I get tremendous satisfaction from learning more and more about the brain although, the amount of material to plow through is huge. I am thrilled to be able to participate and pass on at least some of that knowledge to others – *Caritas sapientis*. (Wisdom and care – Leibnitz)

Of course the more I learned the more I realized that there is a great deal to know and that there is even more that we do not know. The brain is so complex that it will remain a mystery. So what will happen in the next chapters is essentially an organized trip through basic brain physiology and pathology with eight branches of knowledge coming together, often confusing but orderly in its own way, allowing us to make some valid observations and generalizations from the flood of information.

In 1999, when I started *The Healthy Brain Program©*, the idea of brain health, because the brain was an organ of the body, was a new idea. I noted in those early workshops that the field of medicine thought little about the brain and health. That was understandable because until recently we didn't know much about how the brain worked. We had little to signify when it was healthy and we only marginally knew more about what happened when it didn't work i.e. when the brain was sick.

Three things have changed this: First, the relatively new field of medical science called *neuropsychiatry*, driven by powerful investigational techniques, has allowed us to view the brain in real time operation and to understand more than ever before about how it worked. We can now safely take sophisticated pictures of the structure of the brain, measure the brain's reactions and produce revealing images of the brain in action. Neuroscientists can now perform these studies in young and old, in sickness and health, as it happens – safely.

Second, we now know a lot about the major factors that either directly cause brain disorders or set the stage for the emergence of brain disorders later in life. For example, we now know that if you injure your brain when you are young (injuries can be as simple as a concussion or as complex as a disease like meningitis) you are more likely to develop a dementia like Alzheimer's disease later in life. We also know that chronic abuse of alcohol and exposure to other toxic substances make it more likely that psychiatric disorders (which are brain related disorders) and dementia will occur. At a less dramatic level, we know that exposure to chronic and severe stress causes harmful physical changes in the brain. In short, we know with complete certainty that if you don't take care of your brain now you will suffer the consequences – and sooner rather than later!

Third, the neuroscience community has identified many factors that promote healthy brain development from even before being in the uterus as an embryo, to childhood and onward. Those factors protect brain health and build brain reserve, which protect us from brain diseases. As a bonus, a life with a healthy brain is more fun! We now know, for example, that the same exercises that benefit your heart, improve the way that your brain functions. Chapter 9, Physical Exercise, will give you further information on this. We also know that mental and social activities stimulate the brain in very positive ways and that people who engage in an active mental and social life not only live longer and are less likely to develop brain disease than the general population, but report that their life is much more enjoyable. In other words, they add life to their years. This topic is covered in depth in Chapter 10, Mental Activity.

At a very practical level, the brain is our best health care system. Chances are that you have never thought of your brain in this way. That is understandable because most people don't think of the brain at all – unless they have been hit in the head (Chapter 7, Safety) or have failed a test recently (Chapter 10, Mental Activity). The brain has been and continues to be, in fact, the most misunderstood organ of the body. Its purpose in evolution has been to guarantee the survival of our species; therefore, it has been nicknamed the 'biggest sex organ' we have. Hurray! It has succeeded beyond wildest expectations. We now have far more people on earth than we can feed and care about. That has been the brainless outcome of the brain's success without self and social awareness – they too are brain functions.

If we think about the brain, it is mostly about the brain doing what we call thinking. Such a limited perspective sells the brain short. In fact, only part of the brain (the cortex or outer layer) is directly involved in what we call mental activity which includes consciousness and self awareness. What we are aware of as 'thinking' only seems important to the brain because that is the only part of the brain function that we can readily observe. The rest is unconscious. We can give some recognition here to Sigmund Freud and the psychoanalysts although he took the idea from the philosopher Arthur Schopenhauer. Today, neuropsychiatrists can observe conscious and unconscious activity, not as theories, but as real events.

Part I of this book presents an orientation to bring readers up to date on brain science in a general sense and it introduces the important concepts of *Health Span* and *Brain Health*. Part II of this book provides a detailed health care method, *The Healthy Brain Program©*, developed through years of clinical inquiry and experience. It is organized into the *Eight Pillars of Brain Health* and each chapter will cover one of the 'pillars'. You don't need to learn as much as a brain surgeon but you do need to know enough so that the information given in the chapters on the *Eight Pillars* will make sense. For this reason, a glossary of terms has been included for you to refer to for definitions of scientific words used in this book. For those with a deeper interest there are references and a bibliography, but to save space, they are not in the book. You may access those on the website: www. HealthyBrain.org.

Overall, I will avoid using a lot of cross-references and explanatory notes and diagrams. The first edition had too much of that, which made for choppy reading. However, I will keep the reader involved, linking to websites, other books and scientific breakthroughs. I urge you, the reader, to follow through with whatever exercises I recommend – it will be good for the brain!

You might wonder why 'Eight Pillars'. Why not six, seven, ten or twenty? There is a simple answer to this. When I delved into the medical and scientific literature, Eight Pillars covered the primary areas of clinical and scientific inquiry, as piles of papers that emerged from the floor of my study. Each 'Pillar', in fact, stands upon a huge foundation of scientific information, and years of cumulative clinical inquiry. The Healthy Brain Program© strives to be *evidence based.*[1] The knowledge advanced is grounded in a reliable body of scientific research and not merely outdated reports, isolated patient studies and questionable traditional practices. The ideas may be original but the evidence that supports them is built on a solid foundation. This book is the end result of the arduous task of reviewing a vast amount of clinical and scientific literature – not every last bit of data, but care has been taken to ensure that the statements are valid and reliable. The concepts may be new to you, but they are certainly not out of keeping with the opinions of the medical and scientific community.

The best part of evidence-based material is that the sources are available for your own assessment. There is nothing that is hidden, nothing that is magical and nothing that is obscure. In this book there is no 'product' for sale. It is hoped that you will find an understandable and easy-to-follow plan for improving your brain health. By following The Healthy Brain Program©, you will be able to take an important step toward improving the *quality*, not only the *length* of your life. Add life to your years! The Healthy Brain Program© provides a frame of reference that aids in filtering out valid information from *junk science*.[2] My aim is to facilitate your understanding and to improve critical rejection of misinformation, rather than to overwhelm you with voluminous mountains of scientific facts. Keep Albert Einstein's comment in mind:

Not everything that can be counted is important and some things that are important cannot be counted.

Nothing in science, not even science, deserves a free ride. Evidence based medicine can become a straightjacket limiting effectiveness of doctors. What one must remember is that the scientific method is just that – a method. Scientific 'facts' evolve with research. If we lose our healthy skepticism, we will fall into the *science delusion* or 'scientism'[3], an eventuality that I am also trying to avoid.

The *tao* of brain health, as in other aspects of health, is facing the challenge of staying in the middle – avoiding scientism but at the same time avoiding superstition, succumbing to false advertising and ignorance.

Given the above, one will expect that broad empirical tenets of brain health will endure but scientific detail and the depth of understanding will change and evolve. For that reason, an effort will be made to update notes and references which will be found on the website: www.HealthyBrain.org.

PART I

The Brainy Solution to a Long and Healthy Life

CHAPTER 1 ▰▰▰▰▰▰▰▰▰▰▰▰▰▰▰▰

SOME HISTORY AND FACTS

Old age will be postponed so much that men of 60 to 70 years of age will retain their vigour and will not require assistance.

— Elie Metchnikoff, 1903

Every culture up to and including our own has missed the boat on the brain. We know, for example, that the ancient Egyptians thought that the brain was a dirty part of the body. When Egyptians were prepared for mummification, one of the first steps was to quickly remove the brain and dispose of it. The other organs were embalmed and buried with the body. This is quite a dramatic statement in a culture that believed that if the body was disfigured it would have to spend eternity in that state – their version of hell. They could do without the brain but not without the heart, kidneys, liver, hands and feet. Both the ancient Egyptians and their neighbors in Mesopotamia regarded the heart as the primary organ, believing that it housed intelligence and feeling. That notion lasted till the time of Hippocrates. While ancient Greek physicians believed that the brain controlled the mind, their philosopher counterparts believed that the heart controlled both mind and emotion. For a time the philosophers won the argument. Until the 16th Century A.D., most Europeans believed that the heart was the main organ in the body and that the brain had little to do with our thoughts and emotions. (We can thank the above-mentioned Greek philosophers for this misunderstanding). The Bible mentions many of the human organs but never mentions the brain. The first modern neuroscience book was published in the 16th Century A.D., almost 100 years before René Descartes proposed the wild idea that the pineal gland in the brain connected the soul to the body. In the 1860s, Pierre Paul Broca, a French surgeon, was the first person to identify the parts of brain that control speech. In the 1920s scientists invented the electroencephalogram (EEG), the first instrument that was able to measure the electrical activity of the brain. In the 1950s it was shown for the first time that the brain acted as the control centre for the other organs of the body. Progress in the neurosciences has galloped since the 1950s.

1

We can now image the brain in real time, on live people, giving an accurate picture of what parts of the brain are working, and how hard they are working. In the past 20 years we have learned more about the brain than we learned in the previous 5,000 years.

Thousands of years ago biblical scribes recorded that "three score years and ten", or seventy years of age, was a long life, but we know that few people managed to reach that age. In the 17th century life expectancy at birth was only about 25 years. According to folklore, "the man who could not die", St. Germain, who was born in Transylvania in 1694, lived to 90 years. The average life expectancy in the 18th century was 35 years. Fifty was a ripe old age – ninety was forever! By 1899, just over a century ago, life expectancy at birth was 48 years. Dr. Alois Alzheimer, who discovered the disease that bears his name, died of an infection and heart failure at age 51. Times have dramatically changed. People today live longer than at any other point in history and old age has gone from a rare event to a global phenomenon. Alzheimer's disease (AD) and other dementias have grown to epidemic proportions.

Note: From this point and onwards I will try to avoid using the word 'dementia'. Why? The word 'dementia' comes from Latin meaning 'out of the mind'. Its use in medical terminology dates back to the 1700s when 'dementia' had the same meaning as 'insanity'. Because it has become a dirty word; people shun it with shame, which only adds to the stigmatization of the disease. We have enough trouble with stigmatization of psychiatric conditions already – why promote another one? As words like 'crazy', 'stupid', and so forth, 'dementia' is a global derogatory term that has little clinical relevance today. It just scares and shames the hell out of the victim, and that may cause severe anxiety and even suicide. The new DSM-V classification of mental disorders, which tries to be more objective and clinically based, also avoids the term (give nod to the American Psychiatric Association and National Institute of Mental Health). Instead, as doctors we will be using more accurate and less emotionally loaded terms. The new medical term is 'neurocognitive disorder' referring to neuro-degenerative disorders that affect cognition (thinking and memory). We will use the term neurocognitive impairment (NCI) and try, as recommended by the DSM-V, to subclassify the impairment into descriptors that indicate severity and which localize specific parts of the brain. For example, the patient and potential caregivers would not be given a label and told that the victim has 'dementia' but would be advised of specific areas affected by the disease (e.g. mild cerebrovascular disease affecting the frontal areas). Going about diagnosis in a more accurate and rational manner paves the way for understanding, acceptance and the introduction of positive interventions – and that is what this book is all about – *Caritas sapientis*.

Now back to the gist! When we look at *life expectancy* by *age*, in developed nations the trend is clear: life expectancy at birth (LEB) continues to rise, more so for females but males are closing the gap. Caveman had a LEB of only about 20 years – so much for the *Noble Savage*. By the turn of the century LEB in Europe was 50 years and now in the civilized world it is 80 years. LEB continues to rise by about three years per decade. Where will it end, no one knows. Theoretically, mankind, under favorable conditions, could have a maximum life span of 115-120 years.

Today, if you live in North America, which is not the most civilized part of the world, the average LEB is about 75 years. Don't forget that is an average, which means that half of the population lives longer than that. And that may be underestimated. If you manage to survive the perils of early life and live to age 65, it is not unreasonable to expect to live well beyond age 80. As noted, LEB is increasing by three years every decade which translates into the trend that in developed countries, the number of centenarians is increasing at approximately 7% per year. That is a doubling of the centenarian population every decade, pushing it from some 455,000 in 2009 to 4.1 million in 2050.[1] Japan is the country with the highest ratio of centenarians (347 for every 1 million inhabitants in September 2010). Shimane Prefecture had an estimated 743 centenarians per million inhabitants.[2] In the United States, the number of centenarians grew from 32,194 in 1980 to 71,944 in 2010 (232 centenarians per million inhabitants).[3]

What is long life without long health? Let's look at *Health Expectancy*. The difference between *life expectancy* and *healthy life expectancy* can be regarded as an estimate of the number of years a person can expect to live in poor health. Said in a positive way, *healthy life expectancy,* also known as Health Adjusted Life Expectancy (HALE) at birth adds up expectation of life for different healthy states. By definition, it is the average number of years that a person can expect to live in 'full health' by taking into account years lived in less than full health due to disease and/or injury.[4]

Comparison of healthy life expectancy at birth (HALE) and life expectancy at birth (LEB) *

Country	HALE[5]	LEB[6]	YUC **
Japan	76	82	6
Sweden	73	81	8
Switzerland	73	81	8
Canada	72	81	9
United States	70	78	8
Malaysia	61	73	12
Pakistan	56	64	8
India	53	69	16
Sierra Leone	26	41	15

* Figures are rounded off to eliminate decimals
** Subtracting HALE from LEB yields Years Under Care (YUC)

Years in pain and disability are Years Under Care (YUC) – clinical care. That is my term, my inventive way of showing how many years are spent being sick and in relative misery. It is shocking to learn that even in highly developed countries like Japan and Sweden, 6-8 years are typically spent in sickness and misery before death. In many countries it is much more. We have the health technologies to correct that but our deplorable human condition is frustrating.

We have hope. As LEB and aged populations are climbing, so are the number of scientific publications. Studies over the past 20 years have shown that there continues to be a veritable explosion of scientific research and information. About 85% of scientists skilled in scientific methodology were still alive in 1991. At that time, about 24,000 technical and scientific articles were published every day around the world. We have a veritable Information Avalanche.[7]

Doubling the knowledge base*

1750 – 1900	150 years to double
1900 – 1950	50 years to double
1950 – 1960	10 years to double
1960 - 1992	5 years to double

By 2020, information will double every 73 days

* 1992 Conference Teach America, quoted by Gary Starkweather (inventor of the laser printer).

While computer capabilities are doubling every 18 months; 40% of all internet users (USA) are searching for medical information. More than 15,000 websites are devoted to health care. Medical information is doubling every 3.5 years. Our brains have not changed in size for over 200,000 years! But don't lose heart, or shouldn't I say "don't lose your head". The purpose of the following chapters is to distill as much of this information into digestible bytes. We will touch on the peaks of knowledge icebergs – the ride begins. Hold on and do up your seatbelt!

CHAPTER 2

TWO CRITICAL CONCEPTS: HEALTH SPAN AND BRAIN HEALTH

He dies every day who leads a lingering life.

– Pierrard Poullet La Charite, 1590

Health Span

In all likelihood, you are already aware of the need to look after your health. If you are like most of us you see your doctor with some regularity and you are aware of major health risks. More than likely you have heard of the early risk factors and warning signs of heart attacks and cancer. People are generally aware that diet and exercise are important and that medical science has made lots of progress in preventing and treating diseases like stroke, heart disease and cancer. The chances are even better that you have not been introduced to two following critical concepts that can help you to think more clearly about your health and how to keep your health well into old age. The first concept is *health span,* or, very simply, how long you live in reasonably good health. Why is health span important? If you have ever been sick, you know how wonderful it feels to be healthy. If you know anyone who has been chronically sick, you know how devastating poor health can be. Life in sickness is long and full of distress and pain. You need to think about health span because, like life span, it sets an important limit in our lives.

We learned from the previous chapter that your life span is in all likelihood increasing, but is your health span increasing? Without a long health span, we may have a long life that brings years of suffering instead of hale and hardy living. I alluded to this in the previous chapter when referring to YUC, the years under medical care; the time left after healthy years are spent. Living for a long time is no advantage if you are burdened with chronic disease. Even with good medical care, chronic illness creates an unpleasant state of affairs. We could call that 'yucky', as in too much YUC, because it sure is, but simple terms tend to cover up the real-life truth: Chronic illness is fraught with uncertainty, stress, financial strain, lost time, pain and a plethora of complex dynamics driven by negative feelings like grief, dependency, resentment, rejection and distress, and of course, all that is followed by death. Let us now explore the relationship between health span (HALE) and life span (LEB) in three imaginary populations: the *Caveman*, the *Existing* (commonplace) and the *Ideal*.

Caveman had a very short LEB by today's standards. Cavemen had very high mortality rate in the first few years of life, followed by high risks for injury, infection and falling prey to predators. Life was short, brutish and stressful. A thirty year-old caveman was a very old one. On the average, lifespan was only about twenty-five years. HALE was also very brief compared to our experience. Once a caveman got injured or sick, he succumbed very quickly; death was always lurking behind the next tree; if bleeding to death or infection did not kill immediately, the saber toothed tiger surely would. Unfortunately, even today, there are war-torn societies without public health, sanitation, clean water or medical care. Their LEB and HALE are sadly truncated.

In the next scenario we will look at our current situation – the civilized world status quo. That *Existing* model applies to most of us. We can expect to live long; we have a long LEB, over 80 years in some countries. But take another look at the table comparing HALE and LEB in the previous chapter. The YUC factor is big! Depending on what country we live in and what kind of health measures are in place, we can expect to suffer from 6-16 years. Yucky indeed! With some luck we may live a long and healthy life and succumb quickly and relatively painlessly but that is not the norm. Medical science has succeeded in keeping us alive, even when we are very sick, but has failed to instill a yearning for health that could give us an ideal health span. The next scenario will explore what is only attainable by luck even in the existing developed civilizations.

The *Ideal* model of health, nicknamed the "Swedish female model", got its name from the observation made by researchers of population studies, that Swedish women almost achieved the best outcome; long healthy life and short period of terminal illness (long LEB, long HALE and brief YUC). Why were these people different? Research findings were too consistent to attribute the phenomenon to mere chance or a statistical aberration. Since those early research findings, things have only changed a bit. Today, Japanese women show the longest LEB and HALE. If we wanted to keep with the old nomenclature, we could refer to the best attainable pattern (result of genetics and lifestyle), as the 'Japanese female model'. However, to avoid cultural confusion and to keep with the goal, we will stay with the term *Ideal* pattern of living. Please note that this 'ideal' pattern is far from the standard but it is already attainable – with some luck. As many philosophers maintained and as Oprah Winfrey reiterated (BrainyQuote.

com), "I feel that luck is preparation meeting opportunity". We may stack the deck with some lucky cards. The following chapters will prepare you for that but first let's see what the *Ideal* looks like, as noted, in Sweden, other Nordic countries and Japan. The phenomenon is called *morbidity compression*. Most of us live in societies where the situation is somewhere in-between, what we called the *Existing* pattern (UK, USA and Canada). On the average we live a long time but we start ailing soon after middle age. Diabetes, obesity, hypertension and other silent killer diseases are frighteningly common – even amongst children. This is a very uncomfortable situation. We get sick but medical science can keep us going for twenty or more years without the benefit of real health. Some people in this situation are merely inconvenienced but some are in severe pain and require a lot of treatment and support. That is unpleasant and expensive for everybody. Take care of yourself but don't try to live forever.

The ancient Greeks told a story about Aurora, the goddess of Dawn, who became enthused about the vigorous qualities of her lover, Tithonus. She pleaded with her father Zeus to make Tithonus immortal. Unfortunately she forgot to ask for eternal strength and vigour. Because of this lack of foresight, Tithonus was doomed to spend eternity in pain, as a decrepit and pitiful physical wreck. The moral of the story: we should not buy into merely living long; we must aim to live to keep our vitality.

In the present we know three important facts: First, we know with complete certainty that the risk of suffering chronic disease rises as we grow older. Second, we know that most chronic diseases can dramatically deteriorate the quality of a person's life. Third, we know that the way we take care of our health influences health outcomes.

In January, 2009, it was reported in the *Lancet*, a prestigious medical journal, that 44% of all Americans have at least one chronic condition and that 13% have three or more. Most commonly reported chronic conditions in people 65 and over according to the Centers for Disease Control and Prevention[1] were as follows:

Condition	Men %	Women %
Arthritis	45	56
Heart disease	37	26
High blood pressure	54	57
Hearing loss	46	31
Trouble seeing	13	15
No natural teeth	24	25
Depressive symptoms	11	16
Stroke	9	8

Asthma	10	13
Bronchitis or emphysema	10	11
Any cancer	28	21
Diabetes	24	18

It is noteworthy that in the above and most other often-quoted surveys, brain disease causing neurocognitive impairment (NCI) (remember we are using this term to avoid using the word 'dementia') is not mentioned. All of the above except arthritis, lack of teeth and some cancers are major risk factors for brain disease causing NCI. More about this later.

We don't have to go back to ancient days to find out what happens when we neglect our health. When the Soviet Union collapsed, the national health care system began to fall apart. Within a decade two dramatic things happened. LEB began to drop rapidly and HALE went with it in a downward spiral. This was a dramatic turn of events and showed how delicate our balances are. People lived for shorter periods of time and more of that time was spent being miserable, sick and as a burden to their relatives and the health care system.

We also don't have to go far to find an example of what happens when a society takes health span seriously. In Sweden, as mentioned, LEB and HALE figures are very close together because Swedes, particularly the females, not only try to keep from being sick, they actively work at keeping healthy. Swedes, Japanese and citizens of a few other developed countries understand, as few others do, that the best way to treat illness is by staying healthy – by aiming for the *Ideal* model. In Sweden, as one example, people not only live to a ripe old age, but they do so with comparatively low levels of chronic diseases. They live long and live well. Morbidity compression is operating. Chronic diseases, particularly ones that threaten the brain, are 'compressed' i.e. occur for a shorter time and they are squeezed into a much shorter period near the end of the life cycle. By now you have come to understand that we are between the caveman and ideal life outcomes. North American health and life expectancies fall somewhere between the former Soviet Union and the current Sweden or Japan. Our LEB is fairly high but our HALE is not nearly as long as it could be. Look to the chronic disease burden as I outlined above. That is the explanation! Chronic diseases are enemies of the brain.

It is difficult to estimate accurately how many people are affected. When we look at population surveys, the results are sobering. Take a look at the table below which was obtained from a compilation by the Family Caregiver Alliance.[2]

New Diagnoses Annually:	Diagnosis/Cause	Number Diagnosed
	Alzheimer's disease (AD)	250,000
	Amyotrophic Lateral Sclerosis	5,000
	Brain Tumor	33,039
	Epilepsy	135,500
	HIV (AIDS) Dementia	1,196
	Multiple Sclerosis	10,400
	Parkinson's disease	54,927
	Stroke	600,000
	Traumatic Brain Injury	80,000
	Total Estimated Incidence	1,170,062

More than a million people in the USA and about a tenth of that in Canada are diagnosed annually with a chronic brain disorder. That is not counting Huntington's disease, for which figures were not available. Now let's look at how many cases are in actual existence at any given time[2]. That number is known as the *Prevalence* of a disease, in this case, brain disease:

Prevalence:	Diagnosis/Cause	Number Living with Disease
	Alzheimer's disease (AD)	4,000,000
	Amyotrophic Lateral Sclerosis	20,000-30,000
	Brain Tumor	N/A
	Epilepsy	2,000,000
	HIV (AIDS) Dementia	14,537-58,150
	Huntington's disease	30,000
	Multiple Sclerosis	250,000-350,000
	Parkinson's disease	500,000-1,500,000
	Stroke	4,000,000-4,400,000
	Traumatic Brain Injury	2,500,000-3,700,000
	Total Estimate:	13,298,537-16,068,150 people

The above numbers are astounding! They mean that at any given time between 13-16 million people in the USA and over a million people in Canada are living with chronic brain disease. Many of the diseases, particularly Alzheimer's disease (AD), Parkinson's and stroke are age related, increasing with age. Given the aging of our populations, we can expect proportionate increases in the above numbers. The numbers of aged are raising, not only in the Western world, but all over, in Japan, China, Taiwan, India, Europe and so forth.

According to the Alzheimer's Association, at age 65 and over, 1 in 9 seniors has AD and amongst those age 85 or over, 1 in 3 has the disease.[3] Although AD is the most common, there are other brain disorders that cause NCI. Cerebrovascular disease (CVD), is the second most common cause of NCI in the USA and Europe, but it is the most common cause of brain disease in some parts of Asia. The prevalence is 1.5% in Western countries (about 525,000 people in the USA and 52,500 people in Canada) and 2.2% in Japan.[4]

In addition to the above two major brain diseases, AD and CVD, there are some disorders and insults like psychiatric disorders and traumatic brain injury (TBI) which carry their own morbidity to increase the likelihood of NCI. Conservatively, we can estimate that by age 65, the chance of having some form of brain disease is a whopping 8% and that figure doubles every 5 years from ageing as the major risk factor.[5] Other additional risk factors, such as head injury, diabetes, alcoholism, to name a few, will increase risk of NCI. We will discuss risk factors in much more detail in following chapters. Our purpose, of course will be to reduce risk. Scientists who study these matters aim for morbidity compression. Recall that long LEB and HALE approaching LEB increases the time spent in relatively good health with less pain, disability and disease. We all wish to live this way for as long as possible. When there is a *good match* between HALE and LEB, illness comes late and leads to death fast because it comes at the end of the life span. For example, a healthy centenarian will eventually reach frailty and death but at that stage death is usually mercifully rapid. In a *bad match* life scenario, illness starts early but life continues. Medical science can keep people alive for a long time even when they are ill and suffering. For example, someone who has risk factors for stroke (smoking, unfitness, hypertension, etc.) may end up living long, sometimes decades – even with having suffered strokes, pain and severe NCI. Making our HALE and LEB match each other closely should be a goal for each of us. No one wants to outlive ones' healthy years.

I have worked in geriatric care for many years and on each day I have seen the consequences of loss of health including shattered dreams, broken families and much unhappiness. There are few things more distressing than hearing the stories told by our patients of how they had saved and planned for years for trips and family gatherings in their later years and how those plans had been forever lost because they had not been able to maintain a long health span.

The purpose of the above paragraphs was to impress you with the fact that health span, or HALE, are devastated by widespread brain diseases. Next we will discuss brain health.

Brain Health

The second important concept is *brain health*. When I started The Healthy Brain Program© in 2000, the concept was novel. Since then, awareness about brain health (and its converse – the publicity about brain disease) has increased dramatically, but I am still willing to bet that you have never really been introduced to brain health principles. I don't mean that you have not gained knowledge about the brain and brainy matters. Sure you have – it's all over the news media! But do you really *know* what brain health really means? How good it feels? Why would I make such assertions and ask such questions? Because I know with certainty that if you went to your family doctor and asked him or her to start you on a brain health program the answer would be that such things do not exist as part of medical practice. And your doctor would be mostly right. With one or two notable exceptions, such programs do not exist and those that do, are not accessible to the public through the general medical route. The closest thing to a healthy brain program is a healthy heart program, something your doctor is certain to have knowledge of. But you would not qualify even for that unless you already had a heart attack.

I also know, again with complete certainty, that all of us need to consider brain health. You may not be aware of this but we are currently experiencing an epidemic of brain disease. Strokes, AD, Parkinson's disease and other disorders are now among the most common causes of disability and death in people over the age of 65. Attention deficit disorders, developmental disabilities and traumatic brain injuries are leading causes of disability in early life. Depression, anxiety, eating disorders and post-traumatic stress disorders are also very common throughout middle age. It is crystal clear that if we want to maximize our HALE, we have to maximize our brain health. Take note of the following:[5]

- The World Health Organization (WHO) estimates that by the year 2040, AD – representing less than half of all brain disorders – will be the leading cause of death worldwide.
- In 1996 the WHO estimated that by the year 2020 depression, another disorder of the brain, will be the leading cause of disability amongst the working age group worldwide. We are way ahead of that prediction. We are already there!
- In the United States 595,000 people die each year from brain disease (AD and CVD). That is more people than die from either cancer (531,000 per year) or coronary artery disease (490,000 per year).

11

The above information is dramatic and important. If you take care of your heart but neglect your brain, you will be asking for years without health. Heart health does support brain health, especially the vascular insufficiency (cerebrovascular disease) type. We know that risk factors for cardiovascular and cerebrovascular diseases are modifiable; they are not cast into stone. AD is worse in that regard. Although it has a degenerative basis that appears to be more genetic than environmental, some symptoms can be improved or delayed by paying attention to brain health.[6]

Let's look at the two major causes of brain disease causing NCI. (I am proud to report that I am sticking to my rule of not using the term 'dementia'. The true neuropsychiatrists will be happy with me!) The table below shows that the risk for CVD and AD, both increase by age. As we look at specific age groups, the risk accumulates. It is clear that these brain disorders are the most significant challenge to maintaining health in later life. The table below shows us more than that. It shows that if you want to reach old age with a healthy brain you need to start before you are 60. Like heart health, brain health is a life long occupation. Like heart health, if you work at achieving it, your entire life span will be a much better experience for you.

Projected Percentage of Severe Neurocognitive Impairment by Age
Percentage afflicted (Canada)[7]

Age Group	Cerebrovascular d.	Alzheimer's d.	Both
60-64	1	1	2
65-69	1	2	3
70-74	1	4	5
75-79	2	8	10
80-84	5	16	21
85-89	7	32	39
90+	5	64	69

While we may have come as far as understanding that mental function is controlled by the brain, what we have missed, until recently, is that the brain does much more than make us smart or dumb. What the brain does, that is critically important to our health, is to control the rest of our body. Even as late as a few decades ago, the frontal lobes, the main executive functional parts of the brain were misunderstood and referred to as the "silent lobes", implying that their functions were minor and inconsequential. Just the opposite is true!

Let me give you a few facts that will help you to understand what I mean. First, almost one-third of the blood pumped with every beat of the heart goes directly to the brain. Second, the brain uses more energy (25% of the total available), than any other organ system in the body. It seems reasonable to conclude that it is not an accident that an organ with such high energy requirements has some important things to do. What is even more astounding is that this soft, fleshy, bloody thing between your ears, using one-third of your blood and one-quarter of the fuel and oxygen (energy) available to the whole body, weighs slightly less than three pounds.

If you go back to the table above, you will see that in the 85 and older age groups, the figures representing those afflicted are quite high. Yet in the older age groups prediction is difficult. Findings vary according to a range of biases. Some studies have shown that the 85 and older age group of women have lower rates of severe NCI than men.[8] This is a reversal of the trend extrapolated from studies on younger age groups. While some studies suggest that if we live to be 100 our chance of having dementia is 90%[9], other studies suggest that it is only 50%[10]. Some population studies report dementia rates for centenarians as low as 23%.[9] Quite a spread! It is generally accepted that prevalence of AD is about 1% for those just under 65 and this figure doubles every five years.[11]

In reality the brain does quite a few critical things. In fact, as executioners have known for centuries, if you want to quickly kill someone, go for the brain. A bullet to the lower back of the head destroys the part of the brain stem that controls the basic physiological function of breathing and immediately kills the victim. Injuries to other parts of the body are much more survivable. The table below shows some of the functions controlled by the brain:

Some of the Things that the Brain Controls

Perception	Reflexes	Instincts	Emotions	Smell
Sight	Touch	Hearing	Taste	Pain
Coordination	Movement	Hunger	Thirst	Love
Speech	Anger	Personality	Sleep	Arousal
Attention	Cognition	Judgment	Planning	Learning
Memory	Play	Exploration	Muscle tone	Hormones

And too many more to mention.

From here, we can go on to more and more complicated functions. Eating, drinking, sexual activity, body temperature and a host of other basic functions are also controlled by the brain. While pain may start at other points in the body, it ends up being registered and interpreted by the brain. If the brain did not do this, pain would not be felt and appreciated. Just think how dangerous life would be if you had no awareness of injuries, like stepping on a nail or getting your finger caught in a door! Sleep? The brain controls that too. If you want a good night's sleep, don't ingest substances like stimulants that keep the brain from starting the sleep cycle or keeping it going. (We will get to the topic of sleep in more depth in Chapter 11.)

What about the mind? We will come to that later, in the section describing the brain as an organ of the body (Chapter 4) and also in Chapter 10 where we discuss the importance of mental activity.

Rather than go on in this vein let's cut to the chase and talk directly about health. The brain directly influences health because it actively monitors and adjusts biological processes within the body and it works with the immune system to keep disease in check. In short, if you don't keep the brain healthy, you cannot keep the body healthy. If you cannot keep the body healthy you will neither live long nor have a pain free life.

CHAPTER 3 ▬▬▬▬▬▬▬▬▬▬▬▬▬

WHY BRAIN HEALTH AND WHY NOW?

"I maintain that a brain revolution is on the way", I said in 2007, when I wrote the first edition of this book. I was right, the brain revolution has definitely started – everywhere brain health is now up front, in the media; in books, magazines, TV, academic courses and even support groups. With the knowledge that longevity is here to stay and the introduction of the concepts of *health span* and *brain health*, it is time to explore these topics in more detail.

Brain disease is reaching epidemic proportions. As noted, it is already the leading cause of age-adjusted disability worldwide and is projected to become the leading cause of death by 2040.[1] While we have successfully reduced mortality from heart disease and stroke as major killers, we have had little success in reducing the occurrence of almost all brain disorders. The brain is an organ of the body, yet compared to other organs, medicine has little to offer to guide people who are at risk or who are developing brain disease. Physicians and indeed all people must continue to understand our new perspective on the brain – one that goes beyond the minimal, and usually too late, diagnosis of major NCI. We need to do better than the still current medical 'search and disable' approach to people with early brain disorders. Medical practitioners must offer definitive and effective treatment strategies. What's more, we must incorporate already well known principles of health promotion and disease prevention that are highly specific to the brain. Doctors cannot do this because they are focused on treating late stages of disease, not so much on helping in the early stages and generally, they do not practice preventive medicine. Their fee schedules do not permit billing for preventive activities – only for treatment of disease once it is unequivocally present. That is one good reason to address this issue in an educational format. So, we could call The Healthy Brain Program© and the contents of this book as 'applied preventive medicine'. What is wrong with that if it saves lives, pushes back suffering and reduces health care costs? I believe it is the duty of the physician to apply 'wisdom and care' in the treatment of all disorders – *Caritas sapientis*.

In recent years, science has generated well designed, large scale studies of healthy and diseased aged human brains. Sifting this information and translating it into prevention and treatment options is important for care providers and patients alike. We all need fundamentally sound advice and practical approaches to improve brain health. Unfortunately, we are bombarded by conflicting and competing claims about treatments and procedures offering health and longevity. The information explosion is taking place in a disorganized, often even in an anti-intellectual environment, where junk science proliferates – accurate accessible information is often unavailable. Yet, information and more importantly knowledge, can be systematized and made intelligible for the non-scientist reader. It's better late than never.

Let's look more closely into what is happening to us as a society. We have a lot to learn from scientists who study how populations are affected by diseases. They are called epidemiologists. Even more fundamental are studies about population trends affecting healthy populations including population growth and shifts in the composition of large groups of people. Those studies are within the realm of demographics.

We should all be interested in the future because we will all be there! What is really happening? The projected populations of older adults in USA is increasing – is it ever increasing! Let's look at each age group according to a well known study:[2]

- The *'young-old'*, the 65-74 year olds, will increase in numbers so that by 2025 they will reach approximately 30 million (In 1995 they numbered 20 million). Then the *young old* will increase very gradually so the by 2050 they will be at about 35 million.
- The *'old-old'*, the 75-84 year olds, will have increased in numbers to just over 30 million by 2010 and then the numbers shoot up to 60 million by 2030, to level off till 2045 and then drop to about 50 million by 2050. (In 1995 they numbered about 30 million).
- The *'oldest-old'*, the over 85 year olds, will gradually increase to about 80 million by 2030 and will continue to increase to 85 million by 2050. (In 1995 they numbered 35 million).

Clearly, it is the *oldest-old* population group that will generate the greatest concern. Although the above study pertained to the USA, the changes forecast are a worldwide phenomenon. It is happening. Projections by the research organization, *Asian Demographics*, shows China's total population hit 1.3 billion in January, 2005, and will increase to 1.31 billion by 2019 and decline rapidly thereafter. The number of people under age 40 may already have peaked at 800 million and is forecast to decline by 1/3 (over 250 million) over the next 20 years. By contrast, those aged over 40 will grow by 270 million, or more than half, over the same period. Populations aged 60 and older are and will be growing most rapidly. By 2024, they will make up 58% of the population, up from 38% today. By then a full three-quarters of Chinese households may be childless.[3] Other industrialized nations face the same trends. In Japan one in three will be elderly by 2050.[4] In Canada, the number of elderly will increase by 50% while the working age population will increase by just 2%.[5]

Closer to home, in my own province, in British Columbia (Canada), the health care situation is scary. It exemplifies a worldwide trend. Even the Health Minister, was scared by it and publicly admitted that he will focus more closely on his own health. We should do the same. Here is why: According to a report by the *Globe and Mail* (February 15, 2007), we can expect a huge increase in the aged populations. Look at the forecast for the next 25 years below:

Age group	% growth
55-59	38
60-64	82
65-69	137
70-74	133
Over 85	131

"It's pretty easy to do the math. If we are having problems with health care costs now (2007), who will take care of us?" asked the Minister.[6] The number of young people age 20-24 will decline by 8%, while the number of 25-29 year olds is expected to grow by only 9%. The picture gets worse. The tsunami coming at the health care system besides the aging population is *chronic disease* which will multiply costs; diabetes II, limb amputations from diabetes, coronary bypass operations, macular degeneration (primary cause of blindness) – the list goes on! What about the shortage of long term care facilities for the disabled? The above report was 7 years old; the situation has become worse; health care dollars have been cut back. What are you going to do?

We have what the journalists nicknamed a "Geezer Glut" – we went from 'Baby Boom' to 'Geezer Glut' in just a few years. The word *geezer* is a very old one. For thousands of years the Sanskrit verb "to geez" meant "to smile disgustingly while chewing guava", according to Jamie O'Neill, who wrote an article for American Association of Retired Persons (AARP) magazine on the subject. The word also appeared in Chaucer's Canterbury Tales about one thousand years ago and again in Shakespeare's King Lear. The point is that "geezer" (and probably other such terms for the elderly) was not always derogatory but it has come to be used this way to degrade the elderly.[7] This attitude is cultural and is not the norm in non-Western societies. What's more, science is showing us that old brains, as long as they are healthy, function very well and are better and faster at grasping 'the big picture'. Healthy brains continue to grow and become enriched as time goes by. No wonder Jonathan Swift once said, "No wise man ever wished to be younger". I predict that the word 'geezer' or expressions like it will some day be rather complimentary rather than derogatory towards older people.

In summary, the *population pyramid* which used to have a few of the very old folks at the top and a lot of young folks at the bottom is becoming turned upside-down. We are heading into an era where we will have huge numbers of the *oldest-old* at the top and a much smaller number of young people at the bottom. What are the implications? Not enough tax base to finance mass retirement plans. Don't be too confident about the workability of traditional retirement plans – keep working at something you like. It's better for your brain anyway and trends predict that you will need a healthy brain – you may have to work at a second career after retirement.

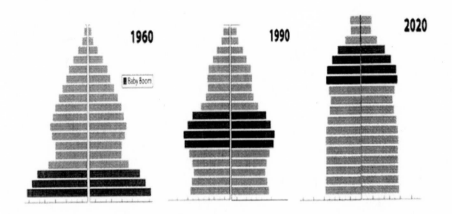

THE INVERSION OF THE POPULATION PYRAMID

The impact of these 'boomers' and the post boom population swell makes demographers and government planners nervous. Phrases like 'Geezer Glut', 'Apocalyptic Demography' and 'Bankruptcy Hypothesis of Aging' plague the pessimists.[8]

Almost daily we get reports of emergency departments filled to capacity, lengthy wait lists for CT scans and surgery and nation-wide shortages of long-term care beds. The effects of longevity are a 'double edged sword'. If longevity fosters long life span without long health span, the burden to caregivers and costs become astronomical. We will prefer to stay autonomous and for that we will need a healthy brain.

Brain health is more important than ever

This is a bold statement. Was it not important for a person to have brain health a hundred years ago? Yes, but the brain did not have to last very long. As described above, we are in the midst of a longevity boom and into a geriatric population explosion. Brain diseases are already the leading cause of disability and death, yet most people believe that cardiovascular disease is the leading cause of death, often quoted at about 37% (1999 – actually even less now), with cancer and stroke following. Cancer is already overtaking heart disease as number one killer. The World Health Organization (WHO) projected that by 2040, AD which represents only about half of illnesses with devastating NCI, will be the leading cause of death. The WHO projected that by 2020, depression will be the second leading cause of disability but major depression is already the primary cause of disability worldwide, years ahead of WHO projections. If you count depressive illness and other psychiatric illness and all neurodegenerative conditions together, brain disease is already the primary cause of disability and death. Today 300,000 Canadians, or one in 13 people over 65, have AD. More than double this number will be affected by 2031. There will be 60,000 new cases of severe NCI each year. (Multiply this by ten for the USA.) In 2001, Dr. Max Cynader, director of the Brain Research Centre at the University of British Columbia, stated, "There is a looming epidemic of brain related illness," and "Brain disease will outstrip heart disease and cancer as the leading cause of death and disability in Canadians in 20 years".

Recent research showed that about 595,000 people die each year in the USA with severe NCI (recall that we are avoiding the term "dementia" – severe neurocognitive impairment (NCI) is dementia). For comparison: 531,000 die with cancer; 490,000 die with ischemic heart disease; 150,000 die with stroke.[9]

The table below is a list of common brain conditions which carry very low mortality rates:

Population Prevalence of Low Mortality Neuropsychiatric Disorders*

Narcolepsy	0.010 - 0.10 %
Tourette Syndrome	0.03 %
Epilepsy	1.7 %
Dyslexia	5 - 10 %
Anxiety Disorders	10 - 15 %
Depression	15 - 25 %**

* Does not include neuroendocrine (hormonal) disorders that are also from the brain and could be conceptualized as brain disorders.

** Lifetime prevalence.

Nobody dies from dyslexia. Nobody dies from narcolepsy unless they fall asleep while driving or have some other accident as a result of falling asleep. Compare this to the list below. These are not as common but they are far from rare. You will note that many of them carry a 100% mortality rate – they are real killers. While they may be slow killers, killers they are. AD may take ten years to kill but nobody can survive it. It is probably the cruelest disease on the planet and some other neuro-degenerative diseases causing NCI are just as bad.

Prevalence of High Mortality Neuropsychiatric Disorders*

Disease	Prevalence	Death Rate
Pick's disease	0.01 %	100%
Huntington's disease	0.02 %	100%
Others	0.03 %	100%
Parkinson's disease	0.20 %	100%
Bipolar disorder	1.0 %***	
Schizophrenia	1.0 %***	
Strokes (per year)	3.0 %	33%
Alzheimer's disease	7.7 %	100%
Depression (major)	15 %***	25% **

* Does not include neuroendocrine disorders.
** Lifetime prevalence.
*** Very high mortality without treatment.

The point of the above is to show you that there are a number of fairly common brain disorders which do not carry a high mortality and people afflicted can lead normal lives. However, there are several, also fairly common brain disorders that involve NCI and will progress to severe disability and death.

Biomedical information explosion

Does it help us to have more and more facts and factoids? Doctors and the public are flooded with information. Surveys have shown that about 66% of Canadian doctors surfed the Web in 1999 and 76% in 2000. Today, many consultants search

Google or medical websites while they are with the patient. That irritates most patients but we must be forgiving, as the information available is so vast that no doctor can be expected to have a full knowledge of it. So, be glad that you doctor will take the time to do a little research on your condition. About 42% of those MDs who were not using the Internet in 1991 planned to do so. The scientific method is very young; 85% of scientists skilled in scientific methodology were still alive in 1991. Even 15 years ago over 24,000 technical and scientific articles were published each day. Over 90% of what we know today about the brain is the result of research over the last 30 years. Biomedical research information is increasing exponentially.[10] Keep in mind that not all information is useful. We know quite a bit about disease states but there has not been enough research done to fully assess what is good for the human brain. Until this comes along, the brain will get no respect. The absence of scientific evidence about what is good for the brain does not dictate that ways and means to improve brain health are nonexistent – we just haven't defined our research priorities appropriately.

The brain is the weakest link in the chain of organ repair and replacement

The brain suffers in silence. Although we only use a part of our maximal brainpower at any given moment, the other parts are active and ready to serve as backup. Quite a bit of the brain can be absent from duty without immediate, observable consequences. This is true of most other organs as well – for example, the heart, lungs, liver and kidneys. This is known as *reserve capacity.* Lesions in the brain are not felt as pain. Unlike for the heart, we do not have 'angina' when the brain has poor blood supply. Small strokes are always silent and even large strokes may be painless. Brain shrinkage and some cell loss is steady but without symptoms. We normally lose about 500 brain cells per hour. Fluctuations in function cannot be easily measured. We have nothing comparable to what is urinalysis to the kidney or enzyme blood tests for other organs. As a result, the brain is ignored while its reserve capacity diminishes. It seems happy to be ignored because, ironically, one of its main functions is to actively ignore and suppress whatever is not immediately relevant to a task at hand or to survival. By the time the brain is aware of its deficits, damage has usually progressed beyond repair. There is nothing on the horizon to parallel cardiac surgery, transplants, joint replacement, renal dialysis or other commonly available life extension measures.

Caregiver burden is huge

The US Census Bureau predicts that by 2025 there will be two 65 year-olds for every teenager.[11] The nuclear family model, in existence for only two generations, is ill equipped to deal with the very young and the frail elderly. Caregiver stress causes psychiatric and physical illness with huge lifestyle and health care dollar losses which are very real but seldom figured into the equation of brain disease costs. This is true collateral damage.[12]

Quality of life issues

Autonomy is not possible without adequate cognitive function, which in turn, is not possible without optimal prefrontal cortical (brain) function. We only need to look at nursing homes to see the quality of life without autonomy. Robert Sapolsky, the foremost authority on contemporary stress research, said: "I can think of no place more stressful than a nursing home."

Prevention is the key to economic management

Depression is estimated to afflict 19 million people in the USA alone, with economic losses conservatively projected at between $30-40 billion a year.[13] Chronic ailments, affecting mostly the elderly, afflict 99 million people (US). This costs $470 billion in direct health care costs. There is another $234 billion in indirect costs related to loss of productivity. 11.6% of gross national product went to health care (1989).[14] It has increased since then.[15] 28% of Medicare funds go to people with less than a year to live. 80% of Americans die in health care institutions.[16] Divide these figures by 10 to get Canadian counterparts. The cost of AD in Canada is $3.9 billion a year. Recently, the CD Howe Institute, a Canadian think tank, advised that the government must set aside $7 billion per year just to meet the upcoming age related increases in health care costs.[17] That has not happened.

David Shenk, in his book *The Forgetting*, said of AD: "Unless something was done to stop the disease it would soon become one of the defining characteristics of civilization, one of the cornerstones of the human experience".

Chapter 4

THE BRAIN AS AN ORGAN OF THE BODY

The brain is the most complex object that we know of.
— Gerald Edelman, Nobel Prize, immunology,
Director of Neurosciences Institute, USA

The brain is truly an amazing organ. Made of 'flesh and blood', it weighs about 3 lb, merely 2% of body weight; soft like jelly, it is held and protected by the bony skull also called the cranium. The brain sits, or almost floats in a blood bath between our ears. It functions very much like a network of modules and computers sitting in a complex chemical soup. It is astoundingly different from a computer because it can alter itself and the soup surrounding it to regulate its environment and behaviours. It can even grow, prune itself and repair itself after injury. Despite its awesome power and complexity, it remains delicate and vulnerable.

The brain can be conceptualized as a computer, with electrical activity, modules, memory banks, connectivity and internal wiring. Medical scientists who are good at mapping and characterizing the brain's circuitry are neurologists, mostly neurosurgeons, often nicknamed the 'hard wire guys'. Modern psychiatrists tend to understand brain function in terms of chemical messengers, hormones and biochemical processes and reactions. They conceptualize us as 'chemical creatures' who may be easily 'tweaked' into harmony by a prescription; often this actually works. Neuropsychiatrists are the synthesizers, the new kids on the block. They are neither neurologists nor psychiatrists in the strict sense – they integrate both fields. Neuropsychiatrists, informed by advanced neuroimaging techniques, have a discipline of their own. They strive toward a synthesis of previous notions and research findings with observations from present day neuroimaging and other state-of-the-art investigative techniques. But let's get back to the brain. It is most fruitful to think of the brain in a dualistic way. Dualistic thinking is helpful in physics too. For example, light behaves as a wave or as a particle; therefore, we had two theories to explain its phenomenology until quantum physics came along. If you don't understand this, you are not alone – your brain is still working well.

The Brain as a Computer

Compared to a computer, the brain works very slowly, switching packets of information at about 20 Hz (20 cycles per second). A computer performs this task at about 1 GHz, or a billion times per second. A typical computer processor chip is about the size of a regular postage stamp and has three layers. It has hundreds of microscopic switching junctions, like transistors, each with a few connections to adjacent junctions. In this manner, many interconnections are made enabling the processor to perform a huge number of simultaneous steps of logical operations. The brain's processor chip is the cortex which is much bigger. Unfurled, it would be about 5 x 5 feet (25 square feet).[1] Estimates range on this topic. Some are really far out there like the one claiming it is the size of a tennis court! It does not look so large, even as 25 sq. ft., because it is folded over and crumpled into the skull as it covers deeper brain structures. In any case, it's quite big and it has six layers of switching junctions called neurons (brain nerve cells). Each neuron makes tens of thousands of connections with adjacent neurons. What's more, the neurons can grow new connections to meet traffic demands – something no transistor can do!

The table below compares brain and computer:

Comparisons between the brain and computers

	Brain	Computer
Weight	3 lbs.	3-7 lbs. (notebook)
Energy consumption (watts)	14	90-250
Power source*	glucose + oxygen	electricity
Speed (HZ = cycles/second)	slow (20)	fast (1 billion)
Cells/transistors	100 billion	1 thousand
Memory (RAM/ROM) bits**	10^{20}	10^{10}

* Glucose equivalent of about 250 *Smarties* (M&Ms) per day.
** Memory RAM and ROM = 2.8 x 10^{20} bits.

We can learn a lot from computer models of the brain, particularly when it comes to learning and healthy aging. Studies in artificial intelligence (AI) make use of neural network models to gain insight into what happens as these networks go through changes that are characteristic of the learning process. At first, changes in connections and the acquisition of new patterns are rapid but learning seems haphazard and mistakes are frequent. As time progresses, complexity increases and learning seems to slow down, while accuracy increases. Accuracy increases with the square root of the number of observations or learning trials. The longer this process goes on, the slower it becomes – the result of the sheer complexity of continuously acquired information and its configuration within the network. On the long run, accuracy increases. Wild unpredictable results diminish in frequency. Reliability increases.[2] Is that not what we see in the performance of the real human brain as it learns and gains experience? The answer is a definite "yes".

The same sort of thing happens in the healthy human brain as it matures and ages. Children play, like neural networks, to extract rules from the environment. Play is full of mistakes but there are usually no disastrous consequences. Play is pure brain enrichment. In youth the brain has to learn quickly, often to ensure survival and to establish safety at the expense of accuracy and reliability. It has to come to grips with the immediate environment – fast. The older but healthy brain is definitely different. It has mastered the immediate environment and it needs to learn slowly in order to generalize in a way that will produce optimal solutions to complex problems, without oscillations or distractions by new stimuli. The observation that in older people, with healthy brains, the speed of learning has diminished is accurate when judged from the perspective of rapid output. It may appear that a lot of information is going in but not many results are forthcoming. "You can't teach an old dog new tricks", as the saying goes. This makes good sense when you consider that older brains are not in pursuit of 'new tricks'. The decrease in learning with advancing age is not disadvantageous. Why must it be this way? The answer is rather simple. Again, consider the child brain. Every experience has a big impact – every new color, every new sound, every new animal, food, plant and so forth. To an older brain, this is just more of the same, business as usual; every stimulus is not new and makes only a small impact on the whole experience. In fact, it is only when the old brain is breaking down that individual stimuli cause big changes in behaviour and oscillations in output! Healthy brains are characterized by stability rather than rapid response to stimuli.

Comparison of healthy young vs. healthy old brain functions

	Young	Old
Number of brain cells	100 billion	99.5 billion at age 100
Number of connections	high (trillions)	very high
Vocabulary	low	higher
Reaction time	fast	slower
Short term memory	maximal	less

Long term memory	limited	maximal
Flexibility	rigid, reflexive	flexible, lateral thinking
Focus	detail, rapid action	big picture, patterns

The Brain as an Endocrine Gland

Like all other endocrine (hormone regulated) glands, the brain is both a source and target of chemical messengers called hormones. Given that it has evolved over tens of thousands of years while its main function was to ensure survival and reproduction, we may actually think of the brain as the biggest sex organ we have. It has succeeded way beyond the bounds of necessity in reproducing mankind. One might even say it has gone out of control in this regard – and ironically that may kill us all! The brain has internal chemical messengers called neurotransmitters which allow neurons (brain cells) to communicate with each other. Modern psychiatry accomplishes almost miraculous effects by tweaking a mere half dozen of these. Serotonin, dopamine, norepinephrine, gamma-aminobutyric acid, aspartate, and acetylcholine are just a few of the neurotransmitters that we know we can modulate in the brain. There are about 200 others and another few hundred neuropeptides that we know exist, but we have not been able to study them in detail and we cannot manipulate them. Just imagine, all of present day biochemical psychiatric treatment is based on tweaking five or six of the hundreds of chemical messengers. An exciting future in the field of brain hormones and neuropeptides is predicted![3]

The Human Prefrontal Cortex is Huge

Because the frontal lobes are not linked to any single, easily defined function, early theories of brain organization denied them any role of consequence. In fact, the frontal lobes used to be known as 'the silent lobes'. The ancient Egyptians believed that its only function was to produce the snot that came out through the nose! That attitude has been with us till fairly recently and to some extent it still survives as most people do not take the brain, as an organ of the body, very seriously. In fact, the frontal lobes of the brain are the most important to civilization as they regulate complex behaviour and even, as we are finding out, many of the inner physiological functions that proceed unconsciously.[4]

In a curious parallel between the evolution of the brain and the evolution of brain science, the interest in the prefrontal cortex was also late in coming. As science gradually began to reveal its secrets, knowledge about it has been growing at an overwhelming pace. The prefrontal cortex (the front part of the frontal lobes, above the eyes, behind the forehead) is the command post and search engine of the whole brain. It has to manage vast amounts of information from inside and outside of the body and make instantaneous determinations about the functions of the body as well as what we experience as consciousness and conscious decisions. It is directly interconnected with every distinct functional unit of the brain. This is only part of the big 'chip' that evolved (arguably first in women) to give mankind the ability to say "no", inhibit impulses, to think, plan and exercise free will. The first layers of cells of the grey matter of the frontal lobes are essential for what are generally considered human qualities such as a sense of time, forethought, planning, insight, self reflection, self knowledge and initiation of planned action. The prefrontal cortex or its analogues account for 29% of the total cortex in humans, 17% in the chimpanzee, 11.5% in the gibbon and the macaque, 8.5% in the lemur, 7% in the dog, and 3.5% in the cat. That is why we experience the monkey as being more 'human' than a cat.

The brain, in all its complexity, functions as a non-linear dynamic system. Such systems are also called 'chaotic systems' – not as logical nor as predictable as we would like. Other *chaotic systems* include the vastly complex weather systems, ocean currents, earthquake producing pressure zones and ecosystems of the world. Scientists have been systematically recording and analyzing these for at least a hundred years with deeper and deeper understanding, but not much has come forth in the way of reliable, predictive information. Look at global warming as an example. Many scientists predicted that it would happen and many remained skeptical; data indicated all kinds of factors – and it continues to do so, yet now that we seem to feel it, it is more believable. The point is that chaotic systems are vast, logical but remain unpredictable. Impacts can be huge – as with the brain!

Don't let these strange terms scare you off! To make things easier, we will now explore the unfathomable brain through the eyes of physicists and biologists.

Brain Energetics: Quantum Physics and Interactionism

The boundary between biology and physics is arbitrary and changing. What used to be Newtonian physics was essentially *linear logical thinking* where A → B → C → D. That is the kind of logic we experience when we read a sentence printed on a page. It is a limitation imposed by writing language linearly on the page and grammatical verbal language itself (not poetry). The dominant side of the brain, usually the left side is the site of verbal logic. The right side is intuitive – the seat of music, art, mathematics (not arithmetic which is verbal). Linear thinking has been imposed not by the natural contours of nature, but by needs to communicate. The pattern tends to persist in the field of biology, and in particular in the field of medicine and cookbooks. Limited thinking engenders limited knowledge and simplistic explanations. Quantum physics and interactionism, the state of the art in physics and mathematics today, tell us that in reality events are determined in a highly complex manner.

Our 50 trillion cells, consisting of 100,000 different proteins, created by 100,000 genes and 20,000 regulator genes and our 3 lb. brains consisting of 100 billion neurons with a theoretical 10^{801} interconnections live as one population communicating through the fatty membrane surrounding each cell. The Newtonian-linear approach is far too simplistic. Instead of a linear $A \to B \to C \to D$, imagine the letters connected every possible way. The number of connections, back and forth, is 20. Now imagine a system with 100,000 items (letters) and all its connections. That was an unfair request – the number is unimaginably high. That is the quantum-holistic way of understanding reality and the brain – the main interpreter of what we perceive as reality.

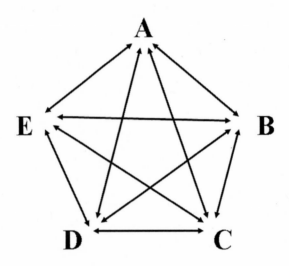

Quantum physics and interactionistic models demonstrate that
events are determined in a very complex manner.

As our knowledge expands, the biological and quantum physical disciplines will merge at certain points, and it is at these points that our understanding of mentation, brain reserve and brain health will rest on a more secure ground.

The brain contains about 100 billion neurons. Those neurons make a trillion connections in each cubic centimeter (the size of a sugar cube). Most of these cells will last your lifetime. The brain loses an estimated 500 cells per hour. Don't despair, cells are renewed and some cells can divide and grow (as we shall see later). Although only about 2-3% of body weight, the brain is a huge consumer of energy. Almost one-third of the blood from each heartbeat goes directly to the brain. Good circulation is the most important function the brain could ask for. Even Hippocrates knew this when he wrote in 500 BCE. "What is good for the heart is generally good for the brain". The heart is a pump that must deliver 75 gallons of blood per hour, one-third of this to the brain. Seventy five gallons per hour translates to 1,800 gallons every day and 657,000 gallons every year. That's enough fresh, oxygenated red blood to fill four Olympic size swimming pools! The heart does this at rest. When physical exertion demands it, cardiac output can increase to six times the resting output.

The brain will get glucose at all costs and the liver is the chemical factory where glucose is made and dumped into the bloodstream. At times of crisis, such as illness or starvation, the liver will cannibalize the other organs and process them to supply glucose. The wisdom of the liver could be stated as, "If I don't supply glucose to the brain, all of us organs will be dead anyway". It will consume the body, burning up first fat, then muscle and other tissues as needed to maintain consciousness.

The brain also uses large amounts of oxygen. Oxygen is carried in the blood in association with iron (hemoglobin). The body cannot make iron; it must be eaten. Vegetable sources of iron are in a chemically bound form and cannot be absorbed and utilized without protein molecules from meat. The only available source of iron is the blood of other animals. We don't drink blood but meat contains lots of blood cells and they contain hemoglobin, the oxygen carrying molecule. More about iron absorption and nutrients in Chapter 8.

Brain size varies widely and intelligence does not correlate well with brain capacity (previously we referred to brain size as being 3 lbs. but that was an average). It correlates better with relative brain size, meaning brain size compared to the body size. So, an elephant and a whale have small brains and small monkeys have big brains. Humans have the biggest brains relative to body size. The heaviest human brain ever recorded was that of a 50 year old male – it weighed 4 lbs. 8.29 oz. reported in 1975, in Florida. The lightest normal, non-atrophied brain on record was that of a 31 year old woman, weighing 2 lbs. 6.7 oz. reported in 1977 in London, UK. We could write volumes on the anatomy and physiology of the brain. We hope that the above basic, knowledge will serve as a platform for understanding the rest of this book, especially when we discuss *The Staying Power of the Brain* in Chapter 6.

The brain builds itself from, and even before birth. It is modular with dedicated parts working with each other. Some parts can compensate for each other. Autopsy pictures of healthy old brains, even 100 year old brains, appear healthy, with minimal atrophy and tissue loss (diseased brains such as AD brains on the other hand, show much thinning of tissue).[4.1]

What about the Mind?

Scientists like to study things they can measure. As Einstein reminded us, not everything that is measurable should be studied – many scientist do so anyway – what a waste of time and tax dollars! And there are phenomena that are difficult to measure but should be studied with more vigor. The study of the mind, as a result, has had a slow start but now, with the advent of non-invasive functional imaging, the mind may be giving up some of its secrets. It better, otherwise, we will have to continue to pretend that it does not exist.

We do not yet know what is mind but we have some ideas – in the mind, of course. From the perspective of this book, what we are looking for is some definitive scientific knowledge about what is good for the mind and what is bad for the mind. We know that mind happens in the brain and it can influence brain function and even endocrine and immune functions. Can we assume that what is good for the mind is good for the brain? Conversely, can we posit that what is bad for the mind is bad for the brain? We have little difficulty in deciding what is good for our car, our dog, our feet or some of our other organs. Healthy Heart Programs address heart disease very well, with predictably and reliably good results.[5] Brain function can affect the mind and mind functions can affect the brain – there is little doubt about that. How do we feed the mind? How do we feed the brain? We will approach those questions again in later chapters on The Healthy Brain Program©; Eight Pillars of Brain Health, in Part II of this book.

How does the mind assert itself through the brain? We do not have enough reliable scientific knowledge to answer questions about what happens when you ask yourself about your brain (As, how is my brain doing?) Or when you ask similar questions of your friend, your parents, your kids. How do you assess all that when the brain feels no pain and 'feeling good', as in euphoria or mania, can actually indicate damage to the brain. Nor do we have much information about what happens when we explore what goes on inside our brains, as in meditation, dreaming and various other states of mind. We are at the threshold of a new era of brain studies that will focus on what goes on inside the brain: our consciousness, altered states and how drugs (and foods) affect our brain – the secrets of the brain. You might call this a new era of transparency! Not everybody likes that idea.

The brain remains mysterious and the cortex even more so. It has no pre-wired memories for language and other behaviours; it must learn everything. What it learns is context dependent and to a large extent state dependent. In other words, what synapses you reinforce depends on what is learned along with environmental factors. Two people learning the same words must have different synaptic constellations to deal with them. Much of what is learned and recalled also depends on what chemicals are acting on the brain. For example, what is learned while on a drug say alcohol, amphetamine or a tranquilizer may not be easily recalled in a sober state but will be accessible when intoxicated, as under initial learning conditions.

There is not enough space on a genome, our basic genetic material, to encode for even simple language learning. The human genome consists of about three billion base pairs (information bits). These four bases are simply four recurring molecules of distinct compounds (you don't have to know why they are called bases). Each single base can be one out of four possible bases, which makes for an informational content of each base of two bits and an information content of the entire genome of six billion bits. Recall that one byte = 8 bits. The information content of the whole human genome therefore is only 750 megabytes (MB). This is about the capacity of a standard compact disk (CD). To give you some sense of how much information we are talking about, it is interesting to note that the music industry was unhappy with the originally proposed format of a 10 cm. CD, which would have held only 550 MB of information. That was not enough to record the full length of Beethoven's Ninth Symphony (74 minutes). The standard was therefore changed to the 12 cm, 680 MB format.[6]

The nucleus of each cell of the human body has an information storage capacity (of base pairs) only slightly larger than that of a commercial CD. It can be calculated that even if we only had ten billion (10^{10}) neurons in the forebrain and even if each one only connected to one thousand (10^3) other neurons, we would have ten trillion (10^{13}) connections. Even if we characterized these connections as a single bit of information (existing or non-existing, instead of on-off or modulated, which is really the case), even then, by a very conservative estimate, we would need 10^{13} bits of genetic information to encode for the creation of these connections. That equals 1.25×10^{12} bytes or 1,250,000 megabytes. Therefore, it is demonstrated that the human genome has only a small fraction of the information needed to program the connections in the forebrain. If we had to rely on the information contained in our genetic material to wire the brain, it would have to be several orders of magnitude smaller than what we have now. We would have microbrains, about the size of a peanut or brains made mostly of empty space between cells, instead of the rich and crowded, ever-growing network of living neuronal connections.

In the previous paragraph we have shown that neurons and their connections grow into configurations without a genetic code. They respond to the environment much more than to any inherited blueprint. The brain is not static, it is dynamic and so is its structure. Form follows function, as it does in other natural and engineered phenomena. (Ever wonder why all airplanes are beautiful and all look so similar?) As the mind functions, so the brain tissue develops – form follows function. The 'hard-wiring' is not as hard as we once thought. Now we speak of the 'bio-hardware'. In contrast to hard-wired computers, our brains may be affected a great deal by what software is running. Our 'software' affects our 'hardware' and vice-versa. Bio-hardware and bio-software are interdependent and inseparable; you cannot eject the diskette or memory stick but you have a lot of choice about what goes in. Oh yes, we do know what is good for the brain in terms of cholesterol, vitamins and alcohol, but what about non-oral gratifications? As Manfred Spitzer noted and posed the question: "What is the right food for thought?"[7] How does the mind know when it is eating garbage? How does it get indigestion? Is there such a thing as 'mental indigestion'? Can it get used to eating garbage? Should it? How much crap can it absorb and manipulate before the crap adversely affects it? Is the converse true? Will gourmet quality mind stuff translate into less stress, good feelings, less disease, better people? Dostoevsky said, "In the end beauty will save the world." I posit that beauty will save the brain and that brain will strive to save the world. Hopefully it's not too late.

No research that we know of is available on the above topic – evidence based science and medicine has no evidence. This is because research tends to be driven by the deficit paradigm – now that it's broken, how do we fix it? It is also driven by the profit motive. What cannot be patented and sold tends to receive less scientific attention. But lack of scientific evidence does not mean that evidence does not exist and lack of scientific knowledge does not stop things from existing. What has not been researched is still out there waiting to be captured in our minds, characterized, measured and applied so that we can live more intelligently. We have little choice but to be wise now. Intelligence is knowing what is good for us as a species. If we fail, we will perish through our own ignorant and dysfunctional practices. There is hope – the brain evolves and grows.

CHAPTER 5 $\blacksquare$

OBSTACLES TO BRAIN HEALTH

Nothing that is vast enters into the life of mortals without a curse.

– Sophocles

My clinical experience has taught me two clear lessons: First, when people encounter brain disease and brain damage they are totally unprepared. Most people have never thought about the brain and how important it is to their lives. Consequently, when something goes wrong with the brain they have no knowledge of what to expect. We continue to be amazed at how often people are surprised to find that depression and sleep disturbances invariably accompany brain injury and that loss of the ability to perform complex tasks should occur when AD or some other disease with cognitive impairment becomes apparent. People simply do not know what the brain does.

This is very different from the situation when other organs of the body become compromised. For example, most people have a good layperson's understanding of the heart and know that it is a muscle, that it pumps blood, that it has valves and so on. The medical community has been providing the general public with information about the heart for at least fifty years. After such a period of time, most people also know that there are things that they can do, ranging from exercise to diet management, to both prevent heart disease and recover from damage to the heart.[1]

In retrospect this lack of knowledge of the brain should not be surprising. The brain – even to medical professionals – is the most misunderstood of the major organs of the body. It is fair to say that most people do not even consider the brain to be an organ of the body in the same way as the heart, the liver or the kidneys. Centuries of misconceptions about the brain coupled with lack of empirical knowledge have left us with a heritage of misinformation about the brain. Most people know that the brain resides inside the skull and that it is the basis of intelligence. Other aspects of the brain remain mysterious. How does it work? How does it relate to the rest of the body?

The second lesson is that the average person has no idea of how to care for the brain. Most people have heard that aspirin can help prevent strokes. Many people have heard that some plants like *gingko biloba* may improve mental functions under some circumstances. But is that true? Most have heard that vitamins are good for you. What people get are broad generalizations – mostly misinformation. As Carl Sagan would have put it, their "baloney detectors" are not turned on! If you ask people how to keep the brain healthy there are few answers. Knowledge of brain care is primitive at best and often consists of an amalgam of old wives tales, misinformation and hopeful thinking or even a kind of optimism of delusional proportions based more on superstition than logic. An exposition of obstacles to brain health, or resistance to developing an active interest in it, will follow.

Lack of basic knowledge

The study of health is neglected in our schools. Some schools do pay attention to heart health and they have been nicknamed *'heart smart'* schools.[2] At least they pay lip service to nutrition and the need for exercise, even as they sell highly sugared and caffeinated drinks and pure starch and salt from their vending machines. I would like to see the development of *brain smart* schools where the basics of safety and brain care are taught. This is not rocket science. Basic knowledge exists but is treated as if it was irrelevant. Lack of basic education leads to ignorance of this topic. That of course, breeds misconceptions, irrational beliefs and superstition. Furthermore, it even leads to prejudice against and stigmatization of people who may in fact need medical help.

Mystification

We have seen from previous pages that the brain is marvelous and astoundingly complex. Of course, nobody can understand it fully. Unlike any other organ, it is a 'complex dynamic system' and that is a difficult concept for almost anyone to grasp. Simply, the brain bamboozles us. What is worse, the brain is constantly uncovering vast amounts of information and knowledge, creating more complexity.

Computer capabilities are doubling every 18 months. Medical knowledge is doubling every 3.5 years. More than 15,000 web sites are currently devoted to health care (2001).[3] Nearly 40% of all Internet users in the USA search for medical information.[4] Half of physicians have a web site (USA).[5] Most remedies and claims for longevity treatment are erroneous and even fraudulent. Profusion of information is not knowledge – judgment is required to make good use of knowledge. People should shy away from bumfuzzlers like snake oil salesmen, hucksters and junk science proponents rife in the marketplace!

Lifestyle and cultural factors

Our shame-based attitude to the brain as an organ has historical roots. The tendency was to blame and scapegoat people with troubled minds. Recall the attitude of the ancient Egyptians! The study of the brain has a dark history full of dogma, irrational fears and violence. The last official, legal witch hunt and burning was only three generations ago – here in the Western world.[6] Even today, in some countries the mentally ill, i.e. those suffering from some forms of brain disease are kept in locks and chains.[7] The development of the prefrontal cortex has been central to civilization; *it is the organ of civilization.*[8] Culturally determined attitudes and habits provide for advantages or disadvantages when it comes to pursuing and maintaining brain health. Our post-modern culture is ridden with an epidemic of brain diseases like AD and other diseases with NCI. We don't know why. Is it the high rate of head injuries from auto accidents and head banging sports? Is it information overload? Is it stress overload? Is it a toxin, a virus, poor nutrition? A combination of the above? Some groups seem to reap protection and we know, from research data, that their brain health is not merely a result of being blessed with good genes. Consider specific populations with better than expected brain health when compared with their general population counterparts:

- People of Okinawa, Japan vs. the rest of Japan[9]
- Seventh-Day Adventists vs. the rest of USA (and Europe)[10]
- Mormons vs. the rest of USA[11]
- Nigerians vs. African-Americans[12]
- Remote Canadian natives vs. urban Canadian natives[13]

Studies have shown that when members of the above groups leave their community and revert to a typical modern diet, their protection vanishes and their health span and longevity diminishes.[14]

Psychological Factors

There is much resistance and denial when it comes to talking, and even thinking about the brain and its problems. It seems that we have been anaesthetized against this awesome, and at times terrifying, topic. Perhaps we need *de-anaesthetists* to help experience the healthy brain and to acknowledge its huge importance in quality of life. We are creatures of habit and resistance to change is a common human quality. On the other hand, avoidance, diversion and denial are active processes acting as defenses against feeling anxiety. Now don't get me wrong, I am not advocating a obsession with health and the brain. That would the equivalent of a hypochondriasis like *cardiac neurosis* of Freudian days. Today psychiatrists would call that an *anxiety disorder* and I don't want you to develop that kind of anxiety and obsession – that would be a 'brain neurosis'. In short, overinvolvement with health and brain health can be ridiculous and even dangerous.[15] Too much pursuit of health and overuse of technologies (including diet) associated with that can make you sick. The aim is increased awareness and action in moderation – using the frontal lobes, of course. Let' get back to other obstacles to brain health.

Physiology

Like all other organs made of 'flesh and blood', the brain has its own peculiar vulnerabilities. Unlike most body structures, the brain feels no pain. A surgeon can cut brain tissue without the perception of pain or even loss of consciousness. Many sharp, penetrating brain injuries do not involve loss of consciousness or much pain. Strokes are not painful – most strokes are silent. Some brain injuries may even be accompanied by euphoria – feeling good while brain tissue dies. This has huge relevance to drug addiction related brain disease. The irony is that the brain backfires; it is so specialized in monitoring everything that it does a poor job of monitoring itself.

There are people who have developed expertise in staying focused, having awareness of the state of their brain. Such awareness and life management in accordance to it can be learned – it is not usually automatic. Yoga, Zen, autohypnosis and other ways of inducing altered states of consciousness in the pursuit of the *alpha state* can be mastered with effort. However, less demanding ways of building that skill exist. You don't have to be a master meditator to benefit from the *relaxation response*[16] or techniques such as *Focusing*[17]. In our North American and European cultures the benefits of meditation have been undervalued. There is no doubt that regular meditation, like physical exercise, is good for the brain. There is a calm sense of well being, not euphoria, that goes with achieving a state of emotional equilibrium. Did you ever wonder why drug addicts turn to meditation and often achieve sobriety? It is never the other way around. People who meditate do not turn to drug addiction!

We will return to meditation, alpha waves and other stress busters in Chapter 12.

Aging

Advancing age certainly makes the brain more vulnerable, but healthy aging avoids brain disease. Neurodegeneration is more than aging. As we age, a hormonal decline (menopause, andropause, somatopause, etc.) takes place, most noticeably between ages of 45-55. Canadian researchers Richard Earle and David Imrie found that during this "decade of vulnerability", the average North American male ages 15.2 years, while the average female ages 18.6 years.[18] Brain hormones and neurotransmitters are also diminished during this period of vulnerability leading to memory problems, brain fog, sleep disorders and depressive illness. On the other hand, the stress hormone, cortisol, remains elevated, wreaking havoc in the brain. Some neuroendocrine decline is normal but exaggerated forms of it are not and should not be accepted as 'normal aging'. We will discuss the effects of hormones on the brain in Chapter 13.

Aging is one factor that we cannot change, but *ageism* is another matter. Many people, including people in the health care field, have erroneous beliefs about what is age associated disease and normal aging. For example, many, including the aged as well, accept that memory problems, sleep disorder, pain, depression and loneliness are acceptable features of aging. Not true! Disease states associated with aging are true diseases that have a diagnosis and treatment. Healthy aging has its own rewards and healthy elders do not wish to be young again. However, ageism is an enemy of healthy aging and a true obstacle to maintaining brain health.[19]

Genetics

The brain, like other organs or parts of the body, can and does sustain a certain amount of 'wear and tear' damage. The mechanisms of repair are governed by a set of brain chemicals which in turn are governed by a set of genes. Thus, the repair apparatus of the brain is partially genetically controlled. The process is similar to the healing process when other parts are injured with some important exceptions. When we cut our skin, for example, a bit of blood and some other fluids ooze out. There is some inflammation and pain and the whole thing clots. It dries and a protective scab is formed. When the scab falls away, fresh, pink, healthy tissue emerges. In the brain it's similar to a point but then trouble begins. When there is a little injury in the brain, say a tiny silent stroke, or a tiny blood vessel rupture or nerve fiber rupture from a concussion, some blood cells and a substance called amyloid, a pink waxy substance, is deposited, possibly to seal off the damage. This is inflamation, but there is no sensation of pain. The amyloid itself is toxic to surrounding cells and kills some. Then, after the healing process and inflammation dies down, the amyloid has to be cleared out. It cannot just fall away and disappear like a scab from the skin. It has to be actively removed by a set of chemical reactions. Scientists believe that it is this process that is the weak link in AD. Our brains have a genetically inherited set of chemicals that modulate the latter processes. Some brains do it better than others. People with high risk for AD have a genetically inherited weakness in some aspects of this process. Some may deposit too much amyloid, some have trouble clearing it, some may have a greater tendency to develop inflammation around the amyloid and so forth. Thus, genetics is an important determinant of AD. Of course, non-genetic, environmental protective factors remain very important, particularly for those with a loading of AD genes. It has been shown that healthy brain practices improve quality of life even for those with AD.[20]

Excesses are brain enemies

Excess food consumption – over-nutrition

Too much food, even good food, leads to obesity and abdominal fat deposition, which in turn, creates a chronic imbalance overloading the pancreatic islet cells to produce more and more insulin. Obesity also overloads the heart because the heart then has to pump blood through a lot of extra tissue. A pound of fat contains over a mile of capillaries. Other changes, promoted by the stress response, concomitantly produce Syndrome X also known as the *Over-nutrition Syndrome*. Syndrome X is one step away from what is known as CHAOS Syndrome: Cardiac disease, Hypertension, Adult diabetes, Obesity and Stroke – very bad for the brain.[21] We will come back to problems of over-nutrition in Chapter 8.

Excess free radicals from lack of regular exercise

Free radicals are good oxygen molecules gone bad.[22] Sometimes they are referred to as 'rogue oxygen species'. These oxygen molecules are hyperactive and without adequate dampening will burn up nearby cells. For example, dopamine containing cells may be affected causing Parkinson's disease. At least 75% of dopamine-containing nerve cells have to die before Parkinson's disease becomes manifest. Other cells in the brain are subject to the same oxidative neurodegenerative process. More about this in Chapters 8 and 9, where we discuss nutrition and physical exercise.

Excess blood glucose (sugar) and insulin

Normally insulin acts as an anabolic hormone transporting glucose into cells. For a number of reasons, mainly obesity and inactivity, we become insulin resistant as we age. The net result is more glucose in the blood; less in the muscles. That leads to less protein building and more abdominal fat deposition – as if the body was getting ready for a period of starvation. The high glucose and fat in the blood accelerates atherosclerosis because it enables fatty molecules to stick to the blood vessel walls.

Excess cortisol from chronic stress and toxins

Allostatic load (a loading of stressors) may be in the form of polypharmacy, substance abuse, physical or psychological stress. Aging itself is a factor in cortisol dysregulation. Cortisol levels are three times higher in the elderly than in young people.[23] High-risk groups are children in stressed families, single moms, post-menopausal females, nursing home residents and the elderly in general. Excess cortisol destroys hippocampal cells in the brain; therefore, we include it as a brain enemy. We will cover the topic of stress in Chapter 12.

The Great Escape

The art of escaping brain disease remains elusive. We cannot change our genetic program (yet) and we cannot stop aging. Fortunately the brain has great staying power. In the next chapter we will explore some of the innate resources of the brain which can enable it to remain in good operational condition, running for over 100 years, and in at least one documented case, for as long as 120 years. (Jean Calment, deceased, 1997.[24])

Jeanne Calment was an amazing woman. She was from Arles, southern France. Her father died at age 93, her mother at 86 – she had longevity genes for sure. She was bicycling at age 100 and she took vigorous daily walks. She broke her hip at age 117 and the following year she quit smoking, reportedly on her doctor's advice, to reduce her risk of osteoporosis. She wanted to live longer! This woman had good brain reserve and will power. Will power, the ability to formulate a plan and stick to it takes a lot of plain 'brain horsepower' which she certainly had – she also had a great sense of irony – all signs of good frontal lobe function. Unfortunately, when she became blind from cataracts at age 120, she refused surgery in the erroneous belief that "it was normal for people of my age to be blind"! Was that not ageism? That set the stage for her demise, for without major natural brain stimulation, as from the visual system, the brain is bound to atrophy. Until then she had no serious cognitive impairment but afterwards she declined rapidly and died at age 122. I wonder how she would have fared if she did have the cataract surgery, restored her vision and went on with her interviews as a celebrity?

CHAPTER 6 ▰▰▰▰▰▰▰▰▰▰▰▰▰▰▰▰▰▰▰▰▰▰▰

THE STAYING POWER
OF THE BRAIN

The increasing numbers of high-functioning centenarian super-seniors attest to the fact that the brain was meant to last. Despite its complexities and vulnerabilities, the brain has great staying power. We will have a "billion shades of grey", as the *Economist* put it, pertaining to the phenomenon of increased greying of the population.[1] There are and will be even more high functioning, tax paying oldsters – the longevity dividend.[2] In the following paragraphs you will learn about features and functions that enable the brain to keep on working, often even when handicapped by silent lesions and even with overt disease.

Further support for the brain's staying power can be found in the Georgia Centenarian Study which reported on 165 individuals between 60 and 100 years old.[3] Parameters of fluid and crystallized intelligence were studied, including acquisition and retrieval of new and familiar information and problem-solving ability. It confirmed that overall performance was slightly lower in older cohorts, but in everyday experiences, where crystallized intelligence is most important, no age-related decline was found. This was especially true for the over 75 age group. Able to take an optimistic, constructive view to life's setbacks, these individuals showed not only greater crystallized and fluid intelligence, but a better quality of life.

Phenomena that account for the staying power of the brain

The neuroglia

It's most fitting to start the section on the staying power of the brain by discussing the neuroglia, or glial cells which are cells between the neuron cells. They do not play a role in signaling; they seem inert. They do act as insulation between nerve cells. The word *glia* means 'glue' in Greek, and the glia acquired this name because when they were discovered, no function could be assigned to them. Scientists assumed that their most likely function was to glue the brain cells together by forming a sticky scaffolding for them to climb on. That was a simplistic conclusion. We now know that the glial cells have several functions. The main activity is not merely physical support but feeding nutrients and growth factors to brain cells. They should have been called 'feeder cells' because they are intimately coupled biochemically to nerve cells. Glial cells increase in numbers with brain stimulation.

Neurogenesis: the birth of new brain cells

New brain cells can generate from precursor stem cells and some through regular cell division. Early research by Clarence Cone, head of the Molecular Biophysics Laboratory at NASA's Langley Research Center, made headlines in 1970, when he demonstrated that cell division was strictly controlled by cell membrane activity.[4] That finding had impact on cancer research but interestingly, he also showed that brain nerve cells that do not normally reproduce on their own, could be induced to divide and multiply under artificial laboratory conditions.[5] So, the myth that brain cells do not divide was shaken once again. We know that in some brain regions there is a lot of active cell division taking place. That is called *neurogenesis*. Brain cells in the dentate gyrus of the hippocampus regenerate in animals and humans, even very old humans. Some cortical cells, cerebellar cells and glial cells can also divide. The hippocampal cells are short lived (few weeks) but their reproduction and growth is stimulated by voluntary (not forced involuntary) exercise, spacious and stimulating environments and the learning of new skills. A serotonin boost, estrogen and lowering the stress hormone, cortisol (if elevated), also increase cell division and growth in the hippocampal area. Now that you know this, the rubric *'grow your own'* can have a positive spin!

Complexity and redundancy

In the 1970s the brain was considered to have at least 10 billion (10^{10}) neurons. The number peaked around age 25. It dropped by 6% in an average lifetime. The possible interconnections, based on those figures was calculated to be an astounding 10^{801}, an unimaginably and unmanageably large number. It is a number greater than the number of particles in the universe. Just for comparison, the number of particles in the observable universe has been estimated at 10^{80}. The number of snowflakes that have ever fallen on earth is estimated at 10^{35}. (Those estimates were from 1991.[6]) More recent figures conservatively estimate that we have about 86 billion neurons. Written in mathematical notation that is 10^{11}. That is still a comprehensible number. Each neuron contains about 1,000 (10^3) branches that form connections with other neurons (synapses). Multiplying this gives us 10^{14} or 100 trillion. Large, but still imaginable and manageable as a number. Each neuron communicates with other neurons by way of a variety of synaptic (switching) states. It is estimated that there are at least 10 different states for a synapse – most likely there are many more. This means that the number of different configurations all the brain cells can make are astronomical. Mathematicians who can calculate combinations, the different possible arrangements in a set, tell us that the number if neuronal combinations could be as high as 10 to the 100 trillionth power. That is $10^{100,000,000,000,000}$ (10 with a trillion zeros after it).[7] Neurons live in huge networks that work in concert when needed but many are relatively inactive at any given time. Huge regions serve as backup systems which can take over whenever demands for multitasking are made. Brain regions work in synergy, like a symphony, and as with a symphony, the overall production cannot be predicted from a simple study of its parts. In other words, the whole seems greater than the sum of its parts.

Brain plasticity

The ability of the brain cells to sprout and branch is called *neuroplasticity*. Neuroplasticity allows the cells to:

- Sprout
- Grow branches
- Make new connections
- Establish new functions (learning)
- Take over old functions no longer active (re-learning)

Every part of the body is mapped onto a specific part of the cortex (outer covering sheet of cells six cells deep) of the brain. If you cut off a finger, the area that represented that finger atrophies and thins out. Conversely, if you learn a new activity, say playing a guitar, the brain areas representing the fingers (and sounds) make many new connections and hypertrophy. This is very much like what happens to muscle, something we have all experienced. When you exercise a muscle it gets bigger and stronger – brain regions respond the same way to increased stimulation. Like muscles, they will also get flabby (atrophy) when unused – use it or lose it!

Self-pruning

During neurogenesis some brain cells can give birth to new ones – entirely new brain cells. It was not until 1998 that this was conclusively demonstrated in humans.[8] Until 1998, scientists and doctors believed that once you had your complement of brain cells you could only lose them. However, brain cells cannot grow and divide, branch and expand indefinitely. Complex and yet poorly understood mechanisms regulate cell loss and even loss of information, which we experience as forgetting. It is important to realize that loss of brain cells, loss of information and even the loss of certain functions are necessary and certainly do not always indicate degeneration or disease.

We tend to think of brain cell loss as a very negative phenomenon, but research on brain maturation suggests otherwise. Most studies tend to focus on brain development in young children, but cognitive psychology tells us that some intellectual abilities continue to mature and grow even after adolescence. MRI scans shows that the structure of the human brain matures substantially between age five and early adulthood. Serial studies of brain development during childhood and adolescence, by repeatedly scanning the same individuals over time, also shows an increase in grey matter (cortical cells) before adolescence, followed by a decline after adolescence. Researchers also confirmed previous reports that white matter (deep fibers connecting different brain areas) increases steadily through development.[9]

Brain development patterns were roughly similar in males and females, although their timing differed, reflecting the timing of puberty. Maturation of various brain regions occurred at significantly different times, giving rise to an uneven schedule characterized by *developmental lags and spurts*. That is in contrast to findings in monkeys and other animals that develop more evenly. Importantly, in humans there is a second phase of synapse overproduction and elimination surrounding adolescence, along the same lines as the better-known *synaptic pruning* changes in babies. Another study showed that grey matter decreased (underwent self-pruning) between adolescence (ages 12 to16) and young adulthood (ages 23 to 30) in several parts of the brain, notably the frontal cortex.[10] Researchers point out an interesting parallel between the structural changes they observe and the psychological maturation of cognitive functions.

Functions such as sensory perception and language development are largely mature by adolescence due to early maturation of the brain areas devoted to those abilities. The frontal cortex controls higher cognitive functions, including emotions, organization of complex tasks and inhibition of inappropriate behaviours. These abilities develop relatively late and are generally considered as signs of maturity. They are not carved in stone, however, and can undergo modification. Therefore, it is not a surprise to find that the brain regions that underlie those functions are among the last to mature. It seems that synaptic growth and elimination are not only common but at times necessary and normal. Does it happen in older age groups? We don't know, but obviously, synapses and brain tissue cannot keep growing *ad infinitum*. The brain sculpts itself!

By the way, Einstein's brain did shrink at the same rate as other brains and it did not have more brain cells but it did have a richer network of connections in the temporal lobe and more glial cells to feed the brain cells.[11]

Neuroprotection

The term *neuroprotection* refers to the processes that protect healthy nerve cells from damage and death. Recall the brief discussion on rogue oxygen species and the need to dampen these. That is what antioxidants do. Natural innate antioxidants from physical exercise and a large variety of exogenous antioxidants from a healthy diet can slow brain degeneration. Hundreds, if not thousands, of varieties of antioxidants work by mopping up and deactivating the rogue oxygen molecules that are too active and threaten to damage delicate intracellular membrane components. If these molecules were not dampened (quenched), they would burn up not only the available fuel (glucose) but also the rest of the surrounding molecules that make up the structure of the cells, including those of the brain. Many of the changes associated with natural aging are the result of relentless oxidation of tissues. There are many substances, both endogenous hormones and exogenous foods, that have antioxidant neuroprotective activity.[12] The best neuroprotection comes from regular physical exercise. We will explore neuroprotection again in chapters on nutrition and physical exercise.

Regulation of catabolic activity

Most hormones are anabolic, meaning body-building, but one in particular is catabolic or body-degrading. Health of tissues, including brain tissue, depends on the balance between anabolic and catabolic activity. The whole process is called metabolism. Anabolism + catabolism = metabolism. Too much catabolism and the tissues will start to lose their protein scaffolding and disintegrate. Cortisol, the stress hormone, is catabolic. High cortisol levels preferentially degrade hippocampal cells in the brain[13] and interestingly, also the insulin producing islet cells in the pancreas.[14] Self-regulation of cortisol puts the brake on neuronal degradation in the hippocampus and other regions. A fuller discussion of the stress hormone, cortisol, and its effects will be revisited in Chapter 12.

The blood-brain barrier (BBB)

The BBB protects the brain by preventing most substances from entering brain tissue; it protects the brain from harmful substances. As determined by electron microscopy, the BBB is formed by the lining of blood vessel capillaries supplying the brain, miles and miles of them, adding up to a huge surface area. Substances in the blood that gain rapid entry into the brain include glucose, the most important source of energy, certain ions that maintain a proper medium for electrical activity and oxygen for cellular respiration. Small fat-soluble molecules, like alcohol can easily pass through the BBB. However, some water-soluble molecules pass into the brain only when actively carried by special proteins in the plasma membrane of the endothelial cells. Examples of these molecules include amino acids, some nutrients and many drugs. Excluded molecules include proteins, toxins, most antibiotics and monoamines, such as the neurotransmitters, – which would cause havoc in the brain. Some unwanted molecules are actively transported out via the endothelial cells. The BBB is protective but it can leak if toxins or physical injury acts upon it.

In summary, the brain's potential to thrive and grow is a function of glial support, neurogenesis, neuroplasticity and neuroprotection.

We have now completed the first part of this book which impressed upon you some history and critical concepts. Also, you now have a fundamental understanding of how the awesome brain functions, what are its enemies and what gives it staying power. Next chapters will explore *The Eight Pillars of Brain Health*. But, not so fast! Have a look at the next page diagram to appreciate how the health factors may interact. Have fun!

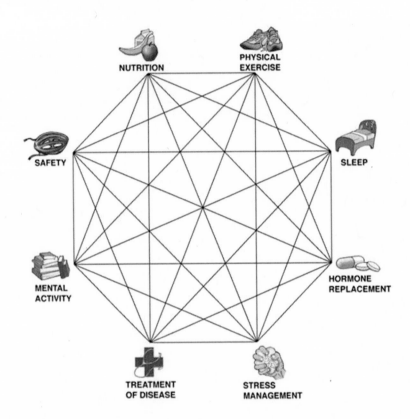

The diagram above depicts how interactions work in the complex dynamics of the real world. No aspect of brain health exists in isolation; all interact with each other. You may find it interesting to imagine specific interactions or find studies that demonstrate such interactions. For example, can you guess how some nutritional effects interact with those of physical exercise? How about interactions between physical and mental exercise? What about the effects of concussion and physical fitness? You can make a game of this and have fun quizzing each other – follow the lines. That will serve as one more brain-building mental exercise. What is more, there is a bonus, a multiplier – the real-life application of what you have thought about – that is learning!

PART II

The Healthy Brain Program: Eight Pillars of Brain Health

CHAPTER 7

SAFETY

With respect to the phenomenology of mild traumatic brain injury, there are believers and nonbelievers. You don't convert until it happens to you or someone close to you.

– William D. Singer, Professor, Neurology and Pediatrics,
Harvard University

I was on my way to my editor's office and while driving along I was making a mental note on how important it was to discuss the chapter on *Safety*. My chapter on traumatic brain injury (TBI), needed revision and I was wondering how I would emphasize the incidence and importance of TBI without becoming too academic and unintelligible to non-medical readers. As I was driving, I saw, up ahead, a body on the road. I pulled over and I was shocked to see a man, face down, nose crushed into the pavement, unconscious in a small pool of blood – a victim of a bike accident. His wife and paramedics were already on the scene and the ambulance had arrived. Fifteen minutes passed and he was still unconscious. When I arrived, late, at the editor's office, I apologetically informed her that I had to take my shoes off because they had human blood on them from the accident. After I described what happened, she confided that she had a son who recently had a skull fracture and concussion in a job-related accident. I did not talk much about the importance of brain safety; the chapter on *Safety,* which I had come to discuss. The fact is that the most important but most difficult aspect of brain health to confront is the topic of brain safety. In this chapter, brain safety refers to the prevention of TBI. This area of brain care is fraught with so much ignorance and danger (to the victim) that it has become ethically, politically, legally and clinically extremely challenging.

Traumatic Brain Injury (TBI)

Bashing a person's head is dangerous – that is a 'no brainer', but milder forms of head injury, particularly closed head injury, remain controversial. Believe it or not, many people, even clinicians, prefer to neglect the whole issue and will even maintain that the brain is quite resilient to trauma. I suspect they are referring to other people's brains, not their own. There are several factors why TBI is under-reported, under-diagnosed and largely ignored – we will cover those points later in this chapter. While the carnage continues, immeasurable suffering and huge monetary cost to the taxpayer goes unchecked. That is why TBI has rightly been called the "silent epidemic".[1] How some people can be so cavalier with the most important organ of the body, the organ of life quality, the organ of civilization, is beyond reasonable comprehension, but this phenomenon of denial not only exists, it is rampant. I hope some facts will change your attitude.

You learned from Chapter 4 – The Brain as an *Organ of the Body*, that the brain is soft in consistency, like jelly, and it sits in a bony cranium with jagged outcroppings and membranes supporting it, thereby keeping it from collapsing into itself. When it comes to acceleration and deceleration forces, this softness and bizarre architecture is precisely the brain's vulnerability.

Perhaps the most shocking aspect of TBI is the sheer extent of the problem. Head injuries are far too common but for complex reasons, emphasis in medical education does not reflect the high prevalence of this disorder. TBI is one of the most common neurological disorders with an incidence of 180 per 100,000. This is equal to the annual incidence of Parkinson's disease, multiple sclerosis, Guillain-Barre syndrome, motor neuron disease and myasthenia gravis combined. About 15% of people who suffered TBI will be permanently symptomatic.[2] Disorders arising from TBI are more numerous than any other neurological disorder with the exception of the common headache.[3] Motor vehicle accidents, falls, assaults and various sports-related activities result in TBI affecting about 7 million individuals, mostly young males, each and every year in North America.[4] Before 1999 in the US, an estimated 2.5 to 6.5 million people lived with the long-term consequences of TBI.[5] Recent reports are more conservative and estimate the casualty at 5.3 million which is a little more than 2% of the whole USA population; that is the number of people who are living with disabilities resulting from TBI.[6] Without a doubt, TBI remains the leading cause of mortality and serious long-term disability among children and young adults.[7]

Also consider the cost to the taxpayer. The price tag of TBI in the USA is estimated to be $48.3 billion annually, second only to AD, which rings in at $115 billion each year. Hospitalized patients account for $31.7 billion and fatal brain injuries cost the nation $16.6 billion each year.[8] Canadian statistics are proportionately similar (divide by 10). The Brain Association of British Columbia serves as an excellent clearing house for related information.[9]

A bump on the head may have serious consequences depending on a number of factors. The type of injury is one factor. Other factors include previous vulnerabilities known as *host factors*. In other words, the end result of a head injury is determined by *who* gets the blow and *how*. While this may seem obvious and common knowledge, understanding the interplay of risk factors is far from common, even among clinicians. There are many types of injury to consider. All other things being equal, a straight head-on blow is not as bad as one with a rotational force. Sharp penetrating wounds that do not accelerate and decelerate the whole brain are remarkably well endured. There are many documented cases of high speed projectiles (small bullets) passing right through the head (cortex) without even loss of consciousness. Even big nails from nail guns have been embedded in heads without loss of consciousness or even awareness of the injury. But a sharp turning force, like a left hook in boxing, or a car accident involving rotation, or a strong body check, will tend to swivel and swirl the brain in its bony case and thereby disrupt brain tissue and brain function.

There is an interesting historical account of a brain injury that highlights the brain's susceptibility and resilience – the case of Phineas Gage. This case was reported by Dr. Harlow in 1868 and it has since been mentioned in most major neurology texts on brain injury. Phineas Gage was a railroad worker who was using a tamping iron to shove a stick of dynamite down a tube used for blasting for creating a railroad bed. Accidentally the dynamite exploded and drove the rod back up right through his head. The sharp end of the rod entered through his face and went through the frontal lobe creating a lobotomy. He did not lose consciousness. Prior to the accident he was a law-abiding, church-going, reliably employed family man but after the accident his personality changed and he became a lazy, slovenly impulsive alcoholic. The history of Phineas Gage illustrates two important observations about traumatic brain injury and the frontal lobes. First, loss of consciousness does not need to take place even when large areas of the frontal lobes are damaged. Second, damage to the frontal lobes results in what are called *personality changes* characterized by apathy, poor impulse control and resultant loss of appropriate social behaviours (aka the frontal lobe syndrome). Recent studies have confirmed that personality disorders commonly develop after TBI and that such changes are independent of injury severity, age at injury or time since injury occurred. Such alterations are illustrative of the effects of both focal and diffuse changes that accompany TBI.[9.1]

Contrary to the popular opinion, that people with head injuries exaggerate their symptoms to get attention or financial gain, victims of TBI with residual deficits tend to lack insight and underestimate and underreport their disabilities.[10] They are collaborators in their own medical neglect.

Cerebral Edema

The *cranium* is the skull, the part of the body that encloses the brain. This large dome of bone is actually made from eight separate bones. They grow and fuse together during early childhood to form the rigid case, the cranium, which protects the brain and sense organs. The back of the cranium is divided into three major depressions, or fossae, in a descending stair-step arrangement from front to back. There are openings in the three fossae for the passage of nerves and blood vessels. The cranium is sometimes called the braincase, but it also houses intimately related sense organs; the eyes, the inner and outer ears for balance and hearing, nasal passages for smelling and breathing and the tongue and palate for taste and swallowing. It is not just the skeletal system for the brain. Babies have 300 bones, but they gradually fuse together, as in the cranium, to reduce the total to 206 fully mature, hardened bones. When the brain is traumatized, like any tissue, it swells. The adult cranium, being one fused bone, cannot accommodate any increase in volume. The infant's brain case, before the fusion of cranial bones, is more forgiving in this regard. A slight tendency to swell can be accommodated but in the adult it is a deadly scenario. The increased pressure from swelling leads to more trauma and swelling. A vicious cycle is set up with the end result of loss of consciousness, convulsions and death. This is particularly likely to happen after a second impact. Unlike for a localized bleed (subdural hematoma) that can be relieved by drilling a hole and letting out the blood, cerebral edema is a generalized swelling and the pressure cannot be relieved by surgical means. Football players have died this way.[11]

Previous and repeated head injury

People who have had a previous brain injury do not recover as well from subsequent injuries. Repeated injury is particularly nasty as deleterious effects are cumulative. The *second impact syndrome* is a well recognized and dreaded phenomenon.[12] In this syndrome, the first injury may not have been severe, perhaps a concussion, but a second one results in cerebral edema and irreversible changes leading to death. Neither the first nor the second impact needs to be particularly severe.

Another aspect of repeated head injury is that each one leaves the victim with a greater vulnerability for another subsequent head injury. After one TBI, the risk for a second injury is three times greater. After the second injury, the risk for a third injury is eight times greater.[13]

We know that direct impact need not take place. Sudden rotation, or whipping to and fro, is as likely to produce TBI. Recent findings demonstrate that neurocognitive abnormalities, as shown by psychological testing, are evident immediately after being dinged even without post-traumatic amnesia (PTA) (confusion) or loss of consciousness (LOC). PTA is more highly correlated with slow recovery than simple LOC.[14]

Host factors are also very important. Who gets the blow? Some brains can take more punishment than others and some brains seem to recover. Next, I will discuss host factors that tend to increase the risk of poor outcome.

Factors that can adversely affect recovery after TBI

At this point, we will not differentiate between *traumatic brain injury* (TBI) and *mild traumatic brain injury* (MTBI), which is otherwise known as concussion. However, it is important to keep in mind that MTBI may have just as serious consequences and the observations that follow apply to both. More about MTBI later.

Genetic status

The presence of apolipoprotein E (ApoE) gene (Alzheimer's gene) predisposes the victims to amyloidosis after all types of brain injury and carriers will have slower recovery and develop delayed consequences after TBI. The mechanism of AD (amyloidosis of the brain) is roughly as follows: Any injury will provoke the healing response and protein like substances accumulate around the dead cells. How the deposition and clearing away of this substance proceeds is genetically predetermined by a set of enzymes regulated by apolipoprotein-E (ApoE) set of genes. The exuded substance is called beta-amyloid precursor protein (beta-APP) which tries to protect the damaged area. In reaction with the dead tissue, it becomes amyloid beta-peptide (A-beta). A-beta, or amyloid, is toxic to surrounding cells causing further damage. Hence, amyloidosis in the brain is a serious matter playing a major role in the degeneration of brain cells after all kinds of injuries: physical injury, cerebrovascular diseases, stroke and AD. It appears that the brain simply cannot get rid of the debris and toxic substances accumulating after cellular injury. People who have inherited a strong tendency to produce amyloid or cannot clear it away efficiently are more prone to Alzheimer's type neurodegenerative disease.

If you find the above too confusing, don't worry about it, just remember that some people have a stronger inherited tendency to suffer chronic deterioration after any brain injury.

Age

The pediatric and geriatric populations are the most vulnerable.[15] Rotational forces are the most dangerous at all ages.

There has been a misconception that infants and children are more resilient to brain injury than adults because their brains are more malleable. In other words, we falsely believed that the developing brain, because of its plasticity, could overcome some destruction. In a very limited way that may be true but when it comes to diffuse injury, infants and children are highly vulnerable. The price to pay? Even after apparent recovery, the injured pediatric population has a higher rate of psychiatric disorders, attention deficit disorder (ADD), hyperkinetic disorder (ADHD), learning disabilities, behavior disorders, addictions and neurocognitive impairment (NCI).[16]

After a TBI, the elderly of both genders are more vulnerable to cognitive deficits and the development of NCI (still avoiding the term 'dementia') following TBI. The risk is also higher for delirium which can be a killer by itself (25% mortality rate[17]) and which often leaves the victim with residual cognitive deficits.[18] The last thing the elderly need is an injury that can impair their capacity to form and recall memories. They are also likely to have a higher risk of falling, leading to repeated injury. Note that older people participate in all kinds of sports but not boxing.

Risk taking behaviours and sports

Athletes in full contact sports such as boxing, football, hockey and even soccer are exposed to head blows or body blows that cause rapid acceleration and deceleration of brain tissue. TBI and especially repeated TBI may result in subdural hematomas, loss of cognitive function or even death.[19] The incidence of TBI clusters around specific high-risk behaviors such as snowmobiling,[20] heading the ball in soccer[21] or in-line skating.[22] We will cover this area in more detail later in this chapter.

Gender

Men, especially young men, are more prone to TBI with immediate or delayed consequences. Men are more likely than females to avoid treatment or minimize the injury. Young people with TBI tend to develop problems in the ADD-ADHD spectrum, while young females may be prone to developing social withdrawal, labile mood and depression – the *quiet type* of ADD.[23]

Other risk factors

Text books also list the following risk factors that may have been present prior to the TBI and which are harbingers of poor prognosis:[24]

- Substance abuse
- Low intellectual functioning
- Previous psychiatric or neurological illness
- Lack of social supports and social networks
- Personality disorder prior to injury
- Suboptimal pre-injury personality adjustment

Now we will explore some specific syndromes.

Shaken baby syndrome (SBS)

Depending on the studies (Canada, the USA, Scotland) reviewed, annual incidence clusters around one in 5,000 children per year. Crude mortality rate in these children is about 20% - 25%. The most common cause of TBI in infants is SBS.[25] Abusive head trauma is implicated in 80% of fatal head injuries in children younger than two years of age. Recent Canadian studies suggest that a minimum of 40 cases of shaken baby syndrome occur in Canada annually with a mortality rate of almost 20% [26] That gives an incidence of one in 4000.

SBS is a term used in medical literature to describe children who have suffered an extremely serious form of child maltreatment that results in a specific constellation of clinical findings and injuries. SBS is similar to a whiplash injury and it may include severe findings such as subdural, subarachnoid and retinal hemorrhages. Even when injuries are not so blatant, damage can occur and escape detection. There may not be any direct blow to the head. Rapid rotational, acceleration and deceleration forces caused by severe shaking alone can result in significant brain injury or even death.[27] SBS occurs primarily in children younger than age three. In most cases they are less than a year old. It is important to note that all that is needed to produce the constellation of symptoms seen in SBS, is for the brain to be subjected to severe enough shaking or rotational force. This mechanism of damage without a direct blow to the head has also been documented in the adult population.[28]

Unfortunately, there are strong reasons to believe that many cases of abusive head trauma often go unrecognized, resulting in further maltreatment and in some cases permanent disability or even death.[29] It is estimated that 30% of cases of abusive head trauma are initially misdiagnosed and the patient is discharged.[30] Caregivers often cite crying as the provoking factor for anger and aggression toward the child. Shaking quiets the child and this perpetuates the behaviour.[31] Shaken babies often develop delayed symptoms which may be nonspecific in nature. They may have failure to thrive, irritability, vomiting, respiratory problems, lethargy or even seizures, coma and death. When doctors examine these babies they find retinal hemorrhages in 50% to 80% of cases. Often there are other forms of abuse with typical telltale signs.

The outcome of SBS is very serious. Only about one-third seem to escape without terrible consequences. One-third suffer permanent injury which may be severe. These may include blindness, seizures and profound mental retardation, spasticity or quadriplegia. Some may have milder symptoms such as developmental delay, mild mental retardation, learning disability and behaviour problems. One-third die.[32]

A study on SBS published in Denver in 1995 reported on 151 such patients. The overall death rate was 23%. Two-thirds of the perpetrators were men, mostly related; in 37% of cases it was the father, in 20% it was a boyfriend and in 13% of the cases it was the mother. What was unexpected was that in 17% of cases the female babysitter was responsible. This is a chilling finding that suggests that even non-abusive parents may end up with a shaken baby.[33]

Of course, the most important cure is prevention, but how do you do that? First of all you can talk about it. Health care professionals should inquire about stress, parenting styles and other causative factors. Parents should be educated and they should be encouraged to educate others. These include caregivers such as babysitters and day-care personnel. Providing resources such as support groups and information web-sites is important. Caregiver parenting skills can be taught. Marriage counseling helps. Substance abuse may be an issue and stress or anger management may be indicated. Poverty is a major stressor in many families and in these cases socioeconomic assistance may help. Parents should be encouraged to ask for help if they feel overwhelmed.

There is little doubt that the SBS is a serious societal problem exacerbated by the pressures of modern life, according to pediatric neurosurgeon, Dr. Norman Guthkletch, who was the first to make the connection between brain swelling/ hemorrhage and the shaking of babies. He admonished that while natural mothers were not likely to shake their babies, casual caregivers were more prone to that behaviour. It was Dr. Guthkletch's opinion that, "…the culture puts a terrible strain on people. It leaves parents unsupported to cope alone with the demands of the economic rat race. Both parents have to work and their children then have to be looked after by an unrelated third party. The grandmother has disappeared because families no longer stay together over the generations". Up until Dr. Guthkletch's work, subdural hematoma was believed to be a complication of skull fracture. The medical establishment in the 1950s was incredulous and could not imagine and believe that this syndrome existed. Dr. Guthkletch actually spoke with a man who admitted to shaking his child. It became clear to him that, "…since the child has a relatively large amount of cerebrospinal fluid around the brain, that the veins bridging from the brain to the skull must be longer in the child. So as the brain goes to and fro in the shaking, the veins get torn, resulting in hemorrhage".

Then came an event that convinced Dr. Guthkletch and others that this syndrome existed. It was described by a former university chairman of neurosurgery who had taken his grandchildren on a roller coaster ride at a fair. A mechanical fault caused the car to come to an abrupt and violent halt. Four days later, the neurosurgeon developed a headache and was able to diagnose himself as having a subdural hematoma. The neurosurgeon wrote about the experience in a medical journal and was recorded as an adult case of *subdural hematoma without direct blow to the head.* "It was the same whiplash injury the infants had suffered," said Dr. Guthkletch. These poor babies with a brain hemorrhage have expanded heads and increased pressure of the cerebrospinal fluid coming from the subdural space. The doctors would drain the fluid and usually the children recovered but then some of them would be back with the same thing a few weeks later because at that stage the doctors weren't conclusive about the fact that it resulted from shaking.[34]

Now we know that you can actually shake the brain stem to pieces. Today, modern imaging technology is the cornerstone for confirming the diagnosis of shaken baby syndrome and radiologists play a key role in identifying cases. This was the message articulated at the First North American Conference on SBS, an event that did not take place till 2004. When a radiologist finds signs of fresh blood along the membrane that divides the intracranial content between the left and right compartments of the brain, it is highly suggestive of SBS. If the radiologist sees acute blood on the babies brain scan and the referring physician says the child was brought in because he or she was drowsy and throwing up, the suspicion is SBS. If we see the same thing in the context of a direct blow to the head, we actually have less reason to suspect that the baby has been shaken. We now know that there is a characteristic appearance, blood in the central membrane, that is specific to this syndrome. This was the discovery of Dr. Jean-Claude Decarie, head of the division of MRI at St. Justine's hospital in Montréal. Dr. Decarie noted, "The accumulation of blood that we see in the scan is a signature of what has happened. But what kills the child is brain damage, and it is not the accumulation of blood that causes the brain damage. It is the shaking that causes the brain damage". Thanks to Dr. Decarie there is now a small crusade being mounted by physicians who work in the field to reduce the incidence of SBS.[35]

The new technology is enabling more diagnoses to be made but it is not the explanation for why the number of cases appears to be going up. It is horrific that people working in the field are of the opinion that overall, there is an increase in the number of cases, rather than that we are just getting better at diagnosing it. They believe that this syndrome, which should not exist at all, is on the rise and the opinion is that SBS is related to the pressures of modern life, where there has been an erosion of moral fabric, disappearance of the extended family and most women having to work, so they don't breast feed and bond like they used to. Thus, there is a link between the stressors of everyday living and traumatic brain injury in infants.

Mild Traumatic Brain Injury (MTBI), also known as Concussion

In discussing MTB, which is concussion, I will use the terms interchangeably throughout the book. When you read "concussion", think MTBI and *vice versa*.

About 80% of TBIs are classified as 'mild' using the Glasgow Coma Scale (score of 13-15) but 10-20% progress to a post-concussion disorder (PCD) – also referred to as post-concussion syndrome (PCS). MTBI is very common, affecting about 100,000 Canadians and ten times as many Americans each year. The most common mechanisms are motor vehicle accidents (45%), falls (30%), occupational accidents (10%), recreational accidents (10%) and assault (5%). One thing is for sure, there is nothing "mild" about mild traumatic brain injury.[36]

Experts quibble over complex definitions but whether an injury is TBI or MTBI is really just a matter of severity. The word concussion comes from the Latin word *concutere* which means *to shake violently*. There have been considerable disagreements among neuropsychiatrists as to how to define concussion. In 1996, the Federal (USA) Traumatic Brain Injury Act gave rise to the early definition. Later, in 2004, the Second International Conference on Concussion classified concussions into *simple* and *complex*.

Simple Concussion

In simple concussion, a person suffers an injury that progressively resolves without complication over 7-10 days. In such cases, apart from limiting playing or training and getting 'brain rest', (more about this later) while having symptoms, no further intervention is required. Neuropsychological screening and testing does not play a role in these circumstances although mental status screening should be a part of the assessment of all concussions in athletes and others.

A simple concussion represents the most common form of head injury and can be appropriately managed by primary care physicians or by certified athletic trainers or coaches who work under medical supervision. The cornerstone of management is rest until all symptoms have resolved; then a graduated program of asymptomatic exertion is prescribed before return to the sport or other physical activity is allowed. All concussions mandate evaluation by a medical doctor.[37]

Complex concussion

Complex concussion, on the other hand, as the name implies, is more serious. This encompasses cases where the victims suffer persistent symptoms including symptom recurrence with exertion and specific negative consequences. These may include convulsions, prolonged loss of consciousness (more than one minute) or prolonged cognitive impairment immediately following the injury. This group may include people, usually athletes, who suffer multiple concussions over time or where repeated concussions occur with progressively less impact force. In this group there may be additional management considerations beyond simple post-concussion management and the usual advice of gradual return to play. Detailed neuropsychological testing and other investigations should be considered in complex concussions. Such patients must be managed in a multidisciplinary manner by physicians with specific expertise in the management of concussive injury.[38]

The Sports Concussion Assessment Tool (SCAT) was developed by a group of neuropsychiatrists and neurosurgeons to help coaches, teachers and athletes in medically evaluating concussion. It is the first and most successful attempt to develop a tool which has become a standardized method of evaluating people after concussion. This of course may be used and should be used to assess people who have had concussion from other causes; therefore, it has also been called the Sideline Concussion Assessment Tool – also SCAT. This tool has been produced as part of the summary and agreement statement of the Second International Symposium on Concussion in Sport that took place in Prague in 2004.[39] The SCAT is available on the Internet and can be downloaded free of charge. See the resources section for the WWW address or just Google "SCAT".

Primary conclusions we have gained about concussion over the past decade:

- The seriousness of outcome and even delayed outcome are a matter of degree of injury.
- A direct blow is not necessary for a concussion to be present.
- Sudden rotation without a direct blow may produce the same effect as closed head injury because of the sudden acceleration and deceleration and twisting of the brain tissue.
- Loss of consciousness is not necessary for MTBI/concussion.
- Studies show that post-traumatic amnesia is a better predictor of later problems than simple brief loss of consciousness.
- Concussion must be made as a diagnosis, graded, and the concussion management protocol must be applied.
- Treatments exist.
- Prevention remains the most important.

MTBI (Concussion) in sports

Soccer

I once spoke to a young girl of about 14 who was proud to be on her school's soccer team. It was the first year that her school had a girls' team and I was curious why she was so keen on the sport. I assumed that competing with the boys gave her a sense of pride and more self confidence. She told me that she loved heading the ball. I asked why? "Because when I head the ball, I see stars," she said! Little did she and her parents know that those little 'stars' were symptoms of brain cells being (at least temporarily) disconnected.

Soccer is a high-risk game but most people are simply not aware of this. There is little doubt that repeated heading of a soccer ball can lead to damage in the brain. Research has shown that heading the ball can lead to cognitive problems like those seen in people with frontal and temporal lobe brain injuries. The Australian neuropsychologist Dr. Rod Markham has reviewed many published studies on this subject and he believes that repeatedly heading soccer balls can cause head and neck trauma as a result of the accumulated effect of small impacts over a period of time. His findings impelled him to write a letter to the world soccer regulator. Dr. Markham states, "I should like to request consideration of the banning of heading of all soccer matches played, due to the cumulative and often acute and long-term brain injuries".[40]

In January 2002, Jeff Astle, a former England international foreward, died at age 59. The coroner ruled that Mr. Astle had suffered neurologic damage due to the repeated heading of soccer balls over his career. This shows what repeated heading can lead to. It is pretty difficult to dispute the coroner's findings. Head injury in soccer is becoming a serious public health issue and the Federation of International Football Associations (FIFA) is aware of it. Players are allowed to wear helmets if they wish and are encouraged to wear headbands with bright coloration with hopes that it will avoid the likelihood of head injury. Of particular concern to us are the children who play the game. As soccer continues to outpace hockey as Canada's fastest-growing children's sport, the issue of heading is a serious problem for minor soccer leagues and players. No one wants children to be at risk and steps need to be taken to ensure brain safety. One of these steps would be to ban heading in soccer altogether for children under 10 years of age. Clearly children do not have strong neck muscles nor the technique to properly head a ball. As for older soccer players, experts advocate the possible use of helmets not only to minimize injuries when hitting balls but also when players collide into one another or hit goalposts. Players should at least have the option of wearing helmets and it shouldn't affect how they play the game.[41] Soccer players, young and old, have a higher rate of NCI than the general population; a Norwegian study showed that soccer players were twice as likely to suffer from NCI, including AD, as members of the general population.[42] Note that a doubling is a 100% increase in incidence!

Football

Football is also dangerous for the brain and the occurrence of MTBI in football has been underestimated. One research project reported that football players were struck in the head 30 to 50 times per game and regularly endured blows similar to those experienced in car crashes.[43] A lead MTBI researcher for Virginia Tech, Dr. Stefan Dumas, Associate Professor of Mechanical Engineering specializing in car crashes and safety equipment, was quoted to say that he did not realize that so many players were absorbing serious hits, especially since only about five had disabling concussions during the season. That study adds to a growing body of evidence suggesting that the incidence of MTBI in football has been underestimated. This means that delayed symptoms of MTBI are being missed or attributed to something else. In one sample, the researchers recorded 3,312 hits during 35 practices in 10 games. Dr. Dumas said, "The interesting part is going to be, these slower speed but higher frequency hits we're seeing on the line. The fact that these players are getting headaches all the time after every game indicates that, there is something going on here". Dr. Dumas used a Head Impact Telemetry System and found that half the hits recorded were greater than 30 Gs. The hardest hits measured more than 130 Gs (a severe car accident = 120 Gs).[44] Better monitoring of football concussions must lead to prevention.

Hockey

The concussion rate suffered by National Hockey League players during the last five years is more than triple that of the last decade. Dr. Charles Tator, from Toronto Western Hospital, and his colleagues, studied injury reports in the Hockey News from the 1990s. After adjusting for number of teams and games per season, the researchers found the rate had soared from four concussions per 1,000 games in the 1986-1987 season to 25 per 1,000 games in the 2001-2002 hockey season. The researchers said faster players, new equipment and harder boards and glass may have all contributed to the increased concussion rate. The study was published in a recent issue of the *Canadian Journal of Neurological Sciences*.[45] The following are some of the factors contributing to head injury in hockey:

- Open ice hits
- Sturdy elbow pads
- Seamless acrylic wall – much stiffer than the older Plexiglass
- Boorish behavior – assaults
- Larger size of players
- Loose chin straps
- Colliding with goalpost
- Helmet to helmet hits

Among hockey players, concussion is common. For example, by the end of December 2000, the NHL had already recorded 67 concussions in the season and the number of seasonal concussions appear to be increasing. Recovering from a concussion is a frustrating test of patience for a fit athlete accustomed to playing with injuries. Dr. Karen Johnston has been a leading researcher in this field. She, as director of a new research unit at Montréal's McGill University Health Center, she said, "…you cannot work through a brain injury. It is a one step forward, two steps back situation." The McGill Concussion Project has been exploring ways to evaluate damage to the brain. As they found, most concussions among athletes go undiagnosed.[46] That is mainly because you don't need to lose consciousness to have a concussion – no LOC needed for MTBI! This is a fairly new concept although it's working its way through the media to the public. Blows to the head have complicated many hockey careers. In 2010-2011, our stellar hockey hero, Sidney Crosby, sustained a concussion that sidelined him for ten and a half months. He continued to suffer from recurrent symptoms but fortunately recovered well enough to play in the 2012-2013 NHL season. When we look at his history, we learn that he did have the wisdom to stay out of the game when he was truly symptomatic and that may have saved his future health. A good example to follow!

Boxing

Boxing legend Muhammad Ali, who was affected by Parkinson's disease, is a case in point. Historically, sports like hockey and football have been studied less systematically. Boxing is unique among sports because it is the only organized athletic activity where inflicting serious brain injury is the goal rather than merely an accepted risk! I came across a Louis Vuitton advertisement in the *The New Yorker* that made me sick.[47] It was called "Some Stars Show the Way" and depicted a healthy looking Muhammad Ali and a little boy, presumably his son, sporting boxing gloves – appearing to be groomed for a career in boxing. I have great respect for athletic achievement and athletic heroes but I draw the line where a direct blow to the head is the object of the game. To popularize that in the name of vanity, fatherhood and male prowess is insane. Would you want your son or daughter to have brain damage at an early age – to be more like you?

When compared to other sports-related injuries, brain injury in boxing tends to be more severe. The *punch-drunk* syndrome was first described as early as 1928 when it was already believed that neurologic consequences from boxing were inevitable. Neurologists have noted that consequences often included confusion, loss of coordination followed by speech disorders and difficulties in motor functioning, including the development of upper body tremors. It was also noted to be associated with Parkinson's disease. It has been estimated that 9-25% of professional boxers ultimately develop the *punch-drunk syndrome*, which also goes by the names of *chronic boxer's encephalopathy, traumatic boxer's encephalopathy* and *dementia pugilistica* (there is that nasty word *dementia* again!) The high prevalence of neurocognitive impairment (NCI) among former boxers, even years after retirement from boxing, has stimulated interest in the relationship between TBI and AD.[48] Traumatic boxer's encephalopathy is associated with increased amyloidosis in the brain suggesting a causative role for repetitive brain injuries in AD. It has been postulated and there is converging scientific evidence suggesting that those who are susceptible to amyloidosis, the carriers of the ApoE risk factor gene, have a higher risk of developing AD after TBI. If you have a family history of AD, don't even think about boxing!

Risk factors - who is at risk? Human post-mortem studies

In humans, increased expression of beta-APP appears to be part of the acute-phase response to neuronal injury (refer to previous explanation of the Alzheimer's gene). Extensive deposition of beta-APP leads to more deposition of beta amyloid protein in the injured brain and the initiation of an Alzheimer's-like disease process within days. These findings have implications for the pathogenesis of AD. The first autopsy study to use brain material to study the connection between TBI and AD, confirmed findings gained from clinical studies.[49] It is well established that deposition of A-beta amyloid plays an important role in AD. In fatal TBI, A-beta amyloid deposition was associated with ApoE gene regulated enzyme activity. TBI induces another related pathology resulting in the formation of microscopic neurofibrillary tangles, another major marker for Alzheimer's type degeneration.

Risk factors - who is at risk? Population Studies

Older epidemiological data support that TBI is a risk factor for subsequent development of AD. In the MIRAGE study, head injury as a risk factor for AD was more strongly associated with subjects completely lacking some ApoE genes (the ApoE gene for the enzyme that slows amyloidosis).[49.1] Saliently, recent longitudinal studies also support significantly increased risk of all types of neurocognitive impairment syndromes and AD in populations with previous TBI. The risk increased as did the severity of TBI. Young adults who experienced moderate or severe head injury were found to have more than double the risk of developing AD and other types of neurocognitive impairments in later life. Any medical history of head injury more than doubled the risk.[50] The worse the head injury, the higher the risk of AD.[51] Moderate head injury yielded a 2.3 fold increase in risk; severe head injury more than quadrupled the risk.[52] More recent population-based studies also suggest that TBI has long term implications.[53]

Risk factors: Animal experimental evidence

Brain researchers have been warning us that boxers, football players, soccer players and hockey players may be at higher risk of developing AD and other forms of NCI because of repeated blows to the head.

Researchers at the University of Pennsylvania conducted their concussion research on mice. One group of rodents was repeatedly bashed in a standardized way. A second group suffered one blow to the head and a third group was not injured. This was also done to mice bred to carry the human Alzheimer's-type CDNR gene, which controls production of a gooey beta protein plaque called amyloid (as we learned of this mechanism previously). The latter breed of animals tended to develop amyloid in their brains similar to that found in people with AD. Scientists know that this buildup of amyloid slowly kills off brain cells robbing sufferers of their memories and ability to function, but they did not know why the buildup begins. The Pennsylvania researchers sacrificed some mice two days after they suffered concussions and others after 16 weeks and measured the amyloid deposits. At each point they saw a dramatic increase of indicators for AD in the mice that received repeated head traumas. (That, by the way, is a very good example of the interplay between genetic and environmental factors.) The above was the first experimental evidence linking head injuries to AD by showing how repetitive concussions can speed up the progress of the disease. The mice were sedated and mechanically subjected to standardized blows that were slightly harder than a slap to the head. Lead researcher, Dr. K. Uruyu said that in a person this would be equivalent to a punch to the head, heading a soccer ball or hitting a helmeted head against a board or another helmet as occurs frequently in hockey and football.[54]

Remarkable advances in brain research have allowed scientists to actually observe what happens in the brain after a concussion. Reported in the journal *Nature* and popularized by the *New York Times*, US researchers at the National Institute of Neurological Disorders and Stroke, a division of the National Institute of Health (NIH), developed a way of looking inside the skull of mice, without disrupting brain function. By shaving microscopic layers of bone off the skull they created a thin, transparent window that allowed direct observation of deeper brain tissue with microscopic lenses, without causing harm. Then they gently compressed the brain to simulate a concussion type injury in which the brain rubs against the hard skull, much like after a football collision or heading a ball. The cameras were able to document, in real time, what took place inside the brain as a result of the injury. What they saw was astounding! The protective membranes around the brain became slightly ripped and frayed by the concussive force and that allowed them to become leaky to molecules usually kept out of uninjured brain tissue (recall the BBB described in Chapter 6). Toxic molecules (free radicals and reactive oxygen species) soon appeared and started an inflammatory reaction in the surrounding, even uninjured brain tissue. Despite the vigorous defensive response of specialized immune cells, the damaging free radicals leaked out too rapidly and caused damage even far from the original impact point. The above findings have provided sobering evidence of how much damage was caused by even minor impacts. Dr. McGavern, the main researcher, saw an opportunity for potential treatment. When given antioxidant drugs immediately after the concussion, almost 70% fewer brain cells died than in the untreated mice.[55] Keep your eye on further NIH research on this exciting topic. We may come up with a post-concussion medication that could reduce damage!

Studies on monkeys have confirmed that acceleration of the head without impact can cause severe diffuse destruction of brain substance.[56] Thus, experimentally, scientists have been able to reproduce the disease process that accelerated NCI after concussions. I know we are not exactly like monkeys but if my monkey cousin can get brain damage from acceleration-deceleration without direct blow to the head, that is enough proof for me to discourage contact sports, particularly in those with a family history of AD or previous TBI.

Post-concussion disorder (PCD)

Of all the people who suffer concussion (100,000/year in Canada), about 20,000 Canadians and 200,000 Americans join the "miserable minority" (not my term but a term used in some neurology texts) each year. These are people who are stuck with Post Concussion Syndrome (PCS) also known as Post Concussion Disorder (PCD) – a tremendous problem for these patients and the physicians trying to help them.[57] PCD encompasses a cluster of symptoms that frequently occur following MTBI. Recognizing the signs and symptoms is the first step in helping these patients with their physical symptoms, cognitive deficits and emotional symptoms. PCD has been described in medical literature for over a century. The term 'post-concussion syndrome' was coined by Strauss and Savitsky in 1934.[58] The constellation of symptoms includes physical symptoms, cognitive deficits and difficulties in emotional functioning. MTBI with or without LOC can cause significant long-term brain dysfunction as detected on neuropsychological testing or PET scanning.[59]

PCD symptoms

Physical: headaches, dizziness or vertigo, unsteadiness or poor coordination, tinnitus (ringing in the ears), hearing loss, blurred vision, diplopia (double vision), convergence insufficiency (trouble with visual focusing), light and noise sensitivity, diminished taste and smell.

Cognitive: impairments in memory (learning or recalling information), attention and concentration, initiation and planning of goal directed activities, judgment and perception, speed of information processing, communication (aphasic and non-aphasic) and increased sensitivity to lack of sleep, fatigue, stress, certain drugs and alcohol.

Emotional: emotional lability, irritability, aggression, personality change, fatigue, decreased energy, anxiety, depression, apathy, disordered sleep, loss of libido, poor appetite or weight gain from lack of self control.

Types of injury

Linear acceleration causes less damage than rotation because the incompressibility of the brain (as any other liquid) limits distortion and disruption of tissue. In contrast, because the brain has so little rigidity in its position in the skull, rotation produces strain and distortion of its moorings. For example, an experienced soccer player can strike a fast-moving ball with his or her head by taking care to avoid rotation of the head on impact. In contrast, an upward and sideways blow to the chin of much less magnitude, by producing a rapid rotation of the hand, may cause loss of consciousness or brain damage. Do you ever wonder why woodpeckers do not suffer concussion?[60] Blows to the back of the head produce a *contre-coup* lesion in the frontal lobes in about 90% of cases, but frontal impact only produces a *contre-coup* lesion in the occipital lobe in approximately 10% of cases.[61]

Coup-contrecoup

I introduced a new term above: *contre-coup*. Let me explain. When the brain is suddenly accelerated and decelerated, as in a blow, or when it hits something or in a whiplash type of injury, it bruises at points of contact with the bony skull. The French expression *coup-contrecoup*, meaning "blow-counterblow" accurately describes the double bruising. The frontal lobes are the most vulnerable. When an injurious blow is received on the forehead, 90% of the time there is bruising of the frontal lobes and 10% of the time of the occipital lobes at the back of the head. Interestingly and very significantly for diagnosis, when the blow is received at the back of the head, the victim still has a 90% chance of a frontal lobe bruising! Diffuse axonal injury (diffuse injury to microscopic nerve fibers) is invisible to the naked eye, and is not usually detected with imaging techniques such as CT scan or MRI scan and may even be unrecognized on autopsy unless microscopic examinations of the underlying cells are undertaken.

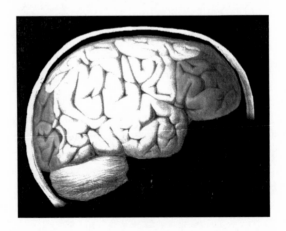

Diagnosis

Traditional medical definitions put too much emphasis on loss of consciousness (LOC) and do not take into consideration the fact that post-traumatic amnesia (PTA) is a more accurate predictor of brain damage. Contrary to official medical definitions such as that in the Diagnostic and Statistical Manual of Mental Disorders (DSM-IV, recently updated to DSM-V) there are many reported cases of patients with long-term PCD symptoms who did not lose consciousness at the time of injury and were therefore not diagnosed.[62] Following concussion people may have impairment in their level of consciousness, but they may not have completely lost consciousness. People, right after a concussion, may also be able to move, talk and respond in a reasonable fashion and yet lack any insight into their actions or retain any memory for events around the time of injury – that is PTA. PTA as an indicator of concussion that may lead to PCD also needs to be adequately understood. The period of PTA is the interval between injury and the time the victim begins to establish *continuous consciousness* (memory) of ongoing events. PTA includes the time during which the victim was awake but confused. People who have had a head injury need to be carefully questioned. Approximately one-third of people with concussion describe an island of recall well before continuous consciousness/continuous memory is re-established, resulting in a potential error in underestimating the total duration of PTA.[63] Clearly, the key concept here is *continuous memory*. Until continuous memory is re-established, the victim is in an amnestic, PTA state. The duration of PTA provides one of the best ways to estimate the extent of organic brain injury, correlating with the severity of diffuse brain damage and the overall degree of disability.[64] Studies show correlation between the duration of PTA and signs of injury, such as duration of time off work, the extent of neurologic disorder and associated physical disability, memory impairment and social and psychiatric disability.[65] Cognitive deficits of PCD are particularly disturbing for victims of head injury because others tend to discredit their reports. A personal account from a neurosurgeon, who experienced PCD following a minor sporting injury, accurately and credibly documented the handicapping posed by the subtle impairment of intellectual functioning.[66]

Checking the family history is very important. If someone is a carrier for the Alzheimer's type gene, which means they are prone to having amyloid depositions secondary to any kind of brain injury, he or she will have a worse outcome. The results of MTBI are difficult to assess and to follow up because denial plays a strong role in coping with the consequences. Victims tend to be ashamed of their symptoms and downplay their disabilities. Presentation and consequences are not uniform. Men in particular, because of their male bravado, tend to deny or minimize events and consequences. It has been shown through sophisticated imaging techniques that axonal injury occurs after traumatic closed head injury, TBI or MTBI, accompanied by cognitive symptoms.[67] New research techniques are beginning to make it clear that brain injury is not static – it is an evolving process. It has been known for some time that MRI scans are more sensitive than CT scans in detecting axonal injury. CT examination is still the standard for life-threatening acute hemorrhage because it is robust and relatively inexpensive. In contrast, while MRI scan is much better for detection of white matter lesions, the exams take longer, are more expensive and require special nonmagnetic ventilators, cardiac monitors and other medical equipment. In this chapter we cannot explore all the different ways a head injury should be investigated and it is important to note that new, exotic techniques are not available in clinical situations – they are used exclusively in research. One thing is for certain, axonal shear corresponding with observed clinical symptoms, without hemorrhage does occur and is difficult to show on routine imaging studies.[68]

Treatment & management

The Brain Injury Association of British Columbia makes the following points in reference to sports-related injury and return to play:

- The player does not have to lose consciousness to have a concussion.
- Symptoms are often subtle.
- It is quite clear from these recommendations that *when in doubt, sit them out!*
- When a player shows any symptoms or signs of a concussion the player should not be allowed to return to play in the current game or practice.
- The player should not be left alone.
- Regular monitoring for deterioration is essential.
- The player should be medically evaluated following concussion.
- Return to play must follow a medically supervised, step-by-step process.

There are guidelines as to when to return to play: medical attention is always advised before any return to play. Concussion management should adhere to the following steps:[69]

1. No activity, complete rest. Once symptoms have cleared, proceed to step two.
2. Light aerobic exercise such as walking or stationary cycling is permitted.
3. Short specific training, for example skating in hockey or running in soccer.
4. Non-contact training drills.

5.   Full contact training after medical clearance.
6.   Return to game or play.

A prolonged recovery and very gradual reentry program is frustrating, particularly for athletes. Concussion largely remains the invisible injury that has no fixed timeline for recovery and, until resolved, keeps the victim anxious with fear of long-term consequences of the impact. The concept of offering treatment and rehabilitation for concussion is relatively new. Given that treatment and rehabilitation from most other injuries is expected by the victim, it is highly surprising that the same attitude to brain injury does not apply! Recall from previous chapters that the brain feels no pain – when there is no pain, the injury is ignored. Most of what people, athletes or otherwise, know about rehabilitation has been learned from experience with pain. An injury is accompanied by pain and as treatment and rehabilitation succeeds, pain diminishes and function returns. Not so for the brain. The concept of rehabilitation is familiar to those caring for brain injured individuals, but it is applied more often after moderate or severe brain trauma rather than following MTBI. Concussion may not be accompanied by pain – it may even involve the opposite – euphoria. It remains difficult to convince people to treat an invisible injury that involves no pain. Athletes have a hard time with this because they are highly motivated to play. There has been an evolution towards management of sports and other head injury; increasingly, we begin timely and actively supervised rehabilitation programs. Rooted in evidence based medical science, workers in the field are transposing this progressive thinking to the management of concussion.[70]

Prevention

Given that prevention is the only cure, wearing protective gear during high-risk activities remains the highest priority. There are some caveats, however. Wearing a helmet can give a false sense of security. It may also lead to faster action and more daring activity. It is very important for young athletes and their parents to note that there are no Canadian safety standards on sport helmets for skateboarding, in-line skating, scooter riding, snowboarding and skiing. There is a campaign led by Richard Kinar, a former professional skier from West Vancouver, to develop safety standards, similar to the Canada Standards Association (CSA) rules for hockey helmets, to improve the effectiveness of helmets in reducing head injuries during skiing, thereby saving quality-of-life, life itself and health-care dollars.[71]

In Canada, the federal government classifies hockey helmets under its hazardous product legislation and requires all helmets sold to meet CSA standards ensuring top-quality products. There is no law requiring other helmets to be CSA approved. Of course, just because there is no uniformity amongst helmets and safety standards, does not mean that they are worthless. Any helmet will offer some protection. Until we have CSA standards, any helmet will be better than no helmet. Studies have shown that provinces that adopt mandatory helmet laws experience about 25% fewer head injuries among children and youths. In 2003, Nova Scotia became the first Canadian province to force in-line skaters, skateboarders and scooter riders to wear helmets.[72]

Helmet Facts:

- Any helmet is better than no helmet.
- When buying a sport helmet, be sure it is appropriate to the sport, fits properly, allows for good visibility and hearing, and has lots of energy absorbing padding.
- Buy from a knowledgeable dealer and watch for international safety standard labels from the American Society of Testing and Materials (ASTM), Consumer Products Safety Commission (CPSC) or the Snell Memorial Foundation.
- When wearing a helmet, ensure it fits snugly and is fastened properly and rides about two finger widths above the eyebrows.
- Unlike hockey helmets, which can be reused after serious hits, most helmets for biking, skateboarding, in-line skating and skiing are known as *single impact* helmets. Once they have absorbed a significant impact they should be thrown away. Do not donate it to someone else!
- Do not buy a second-hand single-impact helmet.
- An examination of admissions to B.C. Children's Hospital emergency department showed that only 20% of injured inline skaters were wearing helmets.
- Broken arms, hands, wrists, fingers are the most common injuries for skateboarders. The incidence of all the head injuries ranges from 12% for skateboarders to 3% for scooter riders. While they occur less frequently, head injuries tend to be more severe than other injuries and can leave permanent damage and impairment.
- Two-thirds of sports injuries involving children and youth occur outside of organized sports.

What happens during a concussion in humans?

We may visualize what happens inside the skull during a concussion. The soft tissue swivels inside the skull and rebounds off the sharp bones. It goes back and forth in there like soft gelatin and it gets bruised and mangled as a result. Many aspects of concussion remain a mystery. It is difficult to extrapolate from animal studies to humans. Doctors do not know how to grade the severity or determine precisely when an injured athlete is fit to play again. Standard CT scans and MRIs are not effective in assessing concussions. Newer high tech studies show that hemorrhage is not necessary to have brain injury because axonal stretching or shear can occur without disruption of blood vessels causing hemorrhage. Traditionally, doctors focused on whether there was LOC. More recent research including the work of the McGill University team suggests that more emphasis should be placed on post-concussion symptoms like headache and even mild amnesia or confusion. As noted, PTA is one of these symptoms which is actually a better predictor of delayed consequences and persistent cognitive impairment than LOC.[73] Other preliminary studies suggest that functional recovery of psychomotor functioning (balance, postural stability) may take a longer time than was previously believed and that PTA, like LOC, was a predictor of delayed recovery. Similarly, recovery of cognitive performance takes a long time after a concussion.[74]

One of the most promising aspects of the McGill work involves measuring electrical activity in the brain. Normally when electrodes are placed on the scalp during an electroencephalogram (EEG) the brain responds to visual and other cues that are programmed from a computer. The brains of concussed athletes who are still experiencing symptoms show abnormal EEG responses but the researchers have found no correlation with the severity of their last concussion. This means that even if the athlete was not knocked out, the brain abnormalities may be greater than someone who was knocked out. Follow-up testing has shown that an athlete's EEG responses returned to normal once the symptoms of concussion have disappeared. This is an important finding because it provides the first means of measuring recovery. In another part of the study athletes perform memory, visual and verbal tasks on a laptop computer while a special type of MRI called a *functional* MRI (fMRI) shows pictures of their brains in action. Athletes who have had concussions frequently complain of specific types of short and long-term memory problems. The test shows researchers if an area of the brain is appropriately activated. If it is, the fMRI will show an increased blood supply to that region. Neuropsychological testing is very sensitive to cognitive dysfunction. It may be wise to undergo studies to establish baseline functioning in people contemplating an athletic career, which may involve head injury. This way performance after a mild head injury could be compared to the baseline and recovery or progression of the condition could be more accurately measured and followed.

Epidemiology of head injury

Often people do not wear their helmets correctly. About 75% of traumatic brain injuries are MTBI and 10-20% progress to PCD. Approximately 45% of head injuries are from motor vehicle accidents, 30% from falls, 10% from occupational accidents, 10% from recreational accidents and 5% from assaults. Two-thirds of these people are males and 50% of them are between the ages of 15 and 34; no medical care is sought by 20-40%. Head injuries account for 80% of bike fatalities. There's a high incidence of head injury in soccer, football and hockey.

As noted, of particular importance is the *second impact syndrome*. Following a first impact with minimal or no symptoms a second impact can lead to death through the mechanism of *cerebral edema*. The risk of delayed symptoms increase exponentially with repeated concussions. It is clear that the more severe the injury the more disability. But even mild head injuries have cumulative effects and are predisposing factors to several serious neuropsychiatric conditions in later life.

PCD has a variable presentation. Majority of people recover in about three to six months. 10-20% have symptoms after six months and that may go on for years. Late consequences can develop. Long-term delayed consequences are associated with early onset NCI, schizophrenia and depressive illness. The syndrome of *persistent post-concussion disorder* (PPCD) can develop. It is a source of enormous suffering and social cost.

The causes of TBI differ by age. In infants the cause is most likely to be SBS, followed by motor vehicle accidents. In young children, falls are the most common followed by motor vehicle accidents and accidents involved as a pedestrian in a motor vehicle accident. In older children motor vehicle accidents are still number one cause and sports are second. In young adults and youth, motor vehicle accidents remain number one, sports injuries are number two and risk-taking behaviour is the third cause of TBI. In the elderly however, falls are more commonly the main cause and motor vehicle accidents are second.

We know a great deal about TBI. Experts can't agree on definitions and diagnosis but there is consensus regarding management. Information comes from sports research, observation of sports injuries, emergency rooms, the SBS, nursing homes and long-term care facilities, safety gear manufacturers and research on animals. Earlier in this chapter we wondered why woodpeckers for example do not develop concussion. The explanation for this appears to be that their acceleration and deceleration of the brain is linear and not rotational and they of course, also have evolved a differently adapted braincase.[75] As we have noted, the brain is very sensitive to rotational especially sudden rotational forces. Even in the neck, sudden rotation can cause damage to the carotid arteries if they have been weakened by atherosclerosis. In one study reported by the Canadian Stroke Consortium, about 81% of neck artery damages leading to stroke were associated with sudden neck movement. About 28% of the cases studied resulted specifically from therapeutic neck manipulation.[76]

Overall, the risk of AD and all other severe NCI syndromes increase exponentially with the severity of the TBI.[77]

THE LEAST YOU NEED TO KNOW

- Head injuries are far too common.
- Concussion can occur without direct head trauma.
- Concussions can have persistent and or delayed complications: depression, apathy, cognitive impairment, early onset neurocognitive impairment (NCI).
- Males tend to minimize the issue – they do not complain.
- Every concussion must be seen by a doctor; insist on a diagnosis and treatment.
- Head injuries can be prevented and prevention is the only cure.
- Education is key.

START NOW!

- Lead by example. If you're a coach wear a helmet.
- Always use protective gear; buckle-up!
- With family history of AD, avoid all contact sports (you and your children).
- Treatment exists. Always see a doctor after head injury.
- In making a diagnosis physicians should ascertain history and high risk activities. Think about concussion as a possible cause of personality, mood and cognitive changes.
- Insist on preventive measures. Talk about it with friends, teammates and family (how common head injuries are, the possible severe consequences, safety issues, helmets, the Shaken Baby Syndrome, what are high risk sports, etc.).
- Educate about common misconceptions and prevention.
- Use the Sideline Concussion Assessment Tool (SCAT). Get your copy of SCAT from the Internet. You are allowed to copy it and distribute it to teachers, coaches and parents. Laminating it is a good idea!

www.HealthyBrain.org

CHAPTER 8 ▰▰▰▰▰▰▰▰▰▰▰▰

NUTRITION

Patient: Doctor, what can I eat to lose weight?
Doctor: That's funny. You automatically think of eating something? There is no
food that can make you lose weight. Stimulant drugs and appetite suppressants
will work for a limited time but they are dangerous. Just eat less, particularly less
junk food and you will lose weight.

We all agree that we need to eat but there are many misconceptions about food
and appetite. In this chapter we will explore some of these misconceptions and
we will introduce principles of good nutrition particularly as it helps the brain.
What we will not do is advance yet one more diet.

When it comes to food we are all strange people living in a strange world. In
some parts of the world there is plenty of food but it's the wrong type. In other
parts of the world people are starving to death. In some parts of the world there are
obese people and people starving to death living on the same block. Sometimes we
eat strange substances or subscribe to fad diets. Much of what we eat is determined
by group pressure, what is easily available or what we have learned as habit.
Eating habits are ingrained early, within the first few years of life.

We live in a very oral society. It is safe to say that in affluent cultures people
hardly ever eat because they need to. We usually eat because of social convention
or as social custom and courtesy and often as entertainment. For example, we offer
food and drink to a guest as a welcoming gesture. We usually do not think about
whether or not the person is hungry or even overweight. Conversely, we accept
food and drink as a friendly response, whether we are hungry or not. Often we
will even insist that a person indulge in eating even if he or she is obese. We may
feel that it would be downright rude not to participate in this kind of exchange.
Eating is so ingrained and interwoven with social existence and harmony that we
expect almost everything to be achieved by taking it in through the mouth – good
feelings, good health, longevity – even weight loss!

Blame the brain – the cerebral cortex! Obesity is a uniquely human problem. Although there may be deep seated mysterious psychodynamics as to why some people eat too much, the widespread phenomenon has a more transparent answer. The cerebral cortex, the most highly evolved part of the brain, has taken over the more primitive part that governs hunger sensation and satiation in animals. It has overruled those centers for so long that we no longer feel hunger. We are programmed to respond to social convention (three meals a day), etiquette, esthetics (visual cues) and habits, which may in fact be addictions. In the socialized human world we feel obligations and cravings – not real hunger. It is well known that 'hunger', is triggered by visual cues rather than the physiological need for food. Our ancestors were different. Like infant humans, they ate small frequent meals, whenever hungry. But not us, we eat when our brain takes an esthetic interest in feeding – when it is triggered by what is visual, appropriate, pleasant and graceful. Hunger signals from the hypothalamus do not get through.

Let us establish a few facts that we can use as a starting point on the subject of nutrition. First of all, in addition to the above harsh introduction, we have to understand what hunger is and how it relates to food intake. Secondly, we will develop some understanding of what nutrition is and what the body needs. And finally we will talk about what good nutrition and bad nutrition mean, particularly as they apply to brain health.

Hunger and cravings

What happens when people don't eat? Well, they get an inner message that they are 'hungry'. But often people get hungry even if they've had a lot to eat only a few hours before. Is this true hunger? We would rather call this at best 'appetite' and at worst 'craving'. We will talk about appetite and cravings and food addiction later in this chapter.

Ordinarily, when a person stops eating they feel more and more hungry until they get some food. The feeling of hunger, when not too severe, is better called appetite. We have an appetite for certain foods after we have not eaten for a while. The more we are deprived of food the less fussy we become and appetite and cravings turn into true hunger – for a few days. Then the body turns to burning ketones and the feeling of hunger disappears. This phenomenon can also be a factor in anorexia nervosa and the apathy and lack of hunger in true starvation.

One misconception is that people in late stages of starvation are actually hungry. People who are starving may not feel hungry! After a period of fasting, if no food is available the body will start to consume itself. The liver starts to produce glucose and when this is depleted from the liver (in a number of days), the body starts to use up first fat and then other tissues. The result, of course, is weight loss and in the end stages of starvation, the weight loss can become so severe and the chemical reactions that go with it so autonomous, that they become irreversible. That is called marasmus, which is deadly. We don't see people with marasmus in North America except in ads appealing for money to help starving children, with their typical muscle wasting, big bellies full of fluid (because of lack of protein to hold the fluid in the tissues) and runny noses. (Kwashiorkor is caused by severe prolonged protein deficiency; marasmus from caloric deficiency). But, until the very end their brains are working. The point is that the body will consume itself to supply the most important organ, the brain, with the fuel that it needs. As noted, after a number of days of not eating anything of caloric value, the body chemistry starts to shift over to producing ketones and the hunger usually triggered by low blood sugar disappears. A state of anorexia prevails. This is why people who have been starving for a long time have to be reintroduced to simple foods in a gradual way. In fact, they may not even wish to eat. They cannot simply sit down and have a big balanced meal without becoming ill. In our society people generally do not feel real hunger. What people feel is an appetite or a craving for a particular food.

Diabetes is an interesting condition because it is actually a state of starvation. The cells cannot adequately utilize glucose. The body (the liver in particular) in its wisdom starts to pour out glucose in an effort to correct that and, as a result, blood sugar level creeps up and more and more insulin is needed to enable the glucose molecules to enter the cells. However, in diabetes the cells are insulin resistant. The end result is a very high blood sugar level despite the muscle and other cells getting inadequate supplies of glucose. Diabetes has been called *'starvation in the midst of plenty'*. We will discuss diabetes in more detail in Chapter 14 when we describe risk factors for brain disorders causing cognitive impairment. Now, back to nutrition and the brain.

It is worthwhile to look at what appetite is and how it is triggered. We live in a society that is raining food upon us and our appetite mechanism (run by the cortex, not as it should be, by the hypothalamus) is triggered far too often. It is not too far-fetched to say that when we eat it is primarily for social reasons and compulsions and not because we need the calories and nutrients. Result: the obesity epidemic that continues to grow. In a world that is raining food, making healthy choices about what and how to eat is not easy. We need some rules to live by.

It is true that human beings have a strong tendency to build fatty tissues. In this respect we are more like the polar bear than any other animal. In a state of nature animals don't get fat even if food is abundantly available and young animals never get fat. Obesity and overweight are only seen in animals that are domesticated or in captivity. Zoo keepers today are well versed in nutritional requirements and are careful to prevent obesity and other forms of malnutrition in animals.

Let's try to understand the psychology and physiology of hunger. Real hunger sense is triggered by the hypothalamus, communicating with the cortex of the brain, giving us the feeling and idea that we want to eat. Particularly in people who are overweight or obese, such signals are not likely to initiate, as they are flooded by sugar and nutrients from overeating, which is staged by the cortex, not the hypothalamus. Feeding is a complex reflex triggered by various stimuli. Once the feeding reflex is triggered, it tends to be autonomous and wants to keep going until it is inhibited by something. The feeding reflex is typically triggered by the sight, smell and taste of food. It is important to understand that the feeding reflex is a relatively simple brain stem motor response, like walking, and like walking, it does not require conscious effort but is still under volitional control. The feeding reflex is facilitated by a number of conditions like an empty stomach, the contractions of the stomach (often experienced as hunger pangs, which they are not), low glucose levels, elevated insulin level, some drugs, environmental cues and psychological factors. The experience of pleasure, meaning the stimulation of the dopamine system of the brain which is the same system that is stimulated by addicting substances, is another great facilitator of the feeding reflex.

The food industry is very aware of what stimulates appetite. For example, in the design of a restaurant a great deal of this kind of knowledge (industrial psychology) is used to create an environment that facilitates the feeding reflex. That includes the ambient music, the size of the servings, the color of the plates and food, the proximity of other people, and so forth. It is probable that a restaurant that serves highly nutritious foods, in the proper amounts that does not promote an entertaining experience and overeating, would not stay in business.

What stops the feeding reflex? While we eat, there is a metering that is registered through to the pharynx that signals to the brain (hypothalamus) that we have eaten. There is also distention of the stomach which gives similar signals. Dehydration tends to inhibit the feeding reflex. Exercise, warm temperatures, some drugs and emotional arousal by some other strong stimulus (for example fear) will tend to inhibit the feeding reflex. There is also a monitoring of the blood by the hypothalamus for various chemicals replenished during feeding. A rising level of blood sugar and a balanced mix of amino acids (from protein) are strong inhibitors of appetite. This is why a hearty broth is more effective in shutting down appetite than sugar and simple starchy foods like potato chips. Of course, plain willpower, a conscious decision to stop eating is very effective but that requires the use of the brain – will-power – in the frontal lobes!

Addicting Substances

Let us digress at this point to introduce *addiction*. There are only about 100 substances that are truly addicting. They act the same way, by stimulating the dopamine mediated pleasure pathways of the brain. That is true for many animal species but we know that it is true for all mammals including man. Truly addicting substances will be chosen in preference to food and sex and will cause death and extinction. Salt, sugar and fat are addicting because they stimulate this dopamine mediated 'reward system'. That explains why some people, particularly those with low self-control like children, people with ADD and people with brain diseases with NCI, find addicting foods irresistible. They will consume them until they are literally stuffed, in preference to foods that have nutrients. As one comedian put it, "I don't eat till I'm full; I eat till I hate myself!"

Body types – genetics, not diet determine body type

Sometimes people confuse body type with nutritional status. We all have a particular body type, or some combination, and we should recognize this and not to try to change it through nutrition. The classification of body types is a science in itself that we will not go into. Simply stated, even 5,000 years ago Sanskrit teachings of Ayurvedic medicine recognized three body types.

Ancient Ayurvedic Body Types

Vata: a thin body type belonging to a person who can be unpredictable and who becomes vivacious and excitable under pressure.
Pitta: a medium body type represents a person who can be intense, orderly, decisive and tends to become angry and abrupt under pressure.
Kapha: a heavy set body type belonging to a relaxed, calm, steady, easy going person who, when stressed may balk and grow silent.

American psychologist William H. Sheldon (1899-1977) classified bodies into three types and also linked personality traits to them:

Sheldon's Body Types

Ectomorph: a person with a tall, thin, fragile frame and a large head with a tendency to be intellectual, introverted, self-conscious and often nervous.
Mesomorph: athletic body types with a muscular, sturdy and thick-necked frame. According to Sheldon, the latter tended to be active and noisy, risk-taking and sometimes insensitive in personal relationships.
Endomorph: rounded, soft, plump bodies thought to be sociable, friendly, relaxed with a fondness for food and comfort.

The above conventional ways of classifying people remain simplistic and in fact we are all combinations – we are each unique. What we do have in common, however, is that, like the polar bear, we have evolved to be very good at building up fatty stores to prepare us for famines. That has been the human condition for at least ten thousand years. It was feast or famine and those who were able to store up a lot of fuel in the form of fat were the survivors. Humans have relatively ten times the fat cells as other animals (except the polar bear). Some native tribes (Pima Indians) have extreme tendency to store fat and also the tendency to develop obesity and diabetes when exposed to a normal North American diet.[1] For 10,000 years we had to cope with famine and every season many died of starvation. We are genetically programmed to store fat. Now we are all suffering from high food pressure from the food industry which produces far more food than we can consume – and they have to sell it! The food industry produces 3,800 calories every day for every man, woman and child in the US. The average person who is not particularly physically active needs about 1200 nutritionist calories (kcal) per day just to maintain weight. The brain uses 25% (300 kcal) of that. A physically active adult man weighing 70 Kg (about 150 lbs) will need about 2,500 calories daily to maintain equilibrium. Our body, like a slow furnace running at our normal body temperature, uses quite a bit of energy.

Let's look at some of the fuel energy involved. The small calorie or gram calorie (cal) is the amount of energy needed to raise the temperature of one gram of water by one degree Celsius. The large calorie, kilogram calorie, dietary calorie, nutritionist's calorie, nutritional calorie, or food calorie (kcal) are all equivalent terms representing the the amount of energy needed to raise the temperature of one kilogram (1,000 grams) of water by one degree Celsius. The large calorie is thus equal to 1,000 small calories or one kilocalorie (kcal). How does that translate into other energy units. The International Standard is the Joule (J). 1 kcal = 4,187 J. The body of that 70 Kg active man, burning 2,500 kcal per day, uses about 2,500 x 4,187 = 10,467,500 J. That is the same as 3 kilowatt-hours of electrical energy, which is about the same as a standard 1,000 watt electrical kettle running for 3 hours. That is equal to 9,927 BTU which is about the same as the energy in 2 KG of TNT – released slowly hopefully.

The food industry

The food industry has a huge problem because they have not only to produce food but they also have to distribute it to millions of people every day. The food has to be appealing enough and safe to eat. That in itself creates a conflict. Food that is fresh and nutritious, free of preservatives, like fresh fruit and vegetables, are delicate; therefore, very expensive to deliver. Processed foods, on the other hand, are cheap to manufacture, have a long shelf life and are easier to store and distribute. Typically, foods that have a long shelf life have to be colorized to give them a fresh appearance and they have to be adulterated by various preservatives, commonly nitrates, to stop them from degrading. Also foods with long shelf life cannot contain vegetable oils such as flax seed oil and olive oil because these oils oxidize (rancidify) quickly. Heavier fats like trans-fats, closer to animal fats, are more stable but unhealthy to eat. Anything that interferes with the bacterial degradation of foodstuff will also interfere with its digestion because digestion is a form of bacterial degradation, like what happens in a composter.

What has all this to do with the brain? A lot. This so-called food pressure, the triggers for feeding and the pleasure we derive from it, are all mediated through our brain. It is our brain that is influenced by subtle and not-so-subtle manipulations to entice us into the pleasures of gastronomy, or should I say gluttony. Unfortunately we pay a big price because we end up eating too much, eating the wrong foods and even eating adulterated and denatured foods. As we shall see, the diseases we incur because of inappropriate nutrition are themselves risk factors for brain disorders with NCI.

Attitude toward food and feeding may have a lot to do with what we become. The prevalence of obesity amongst dogs and cats in the USA is a whopping 25%. Their owners must be in denial because 50% of them do not admit that their pet has a weight problem.[2] It's not just *what* we eat but *how* we eat and *how much* we eat that counts. In the first part of this book, when we were discussing longevity, I mentioned the people of Okinawa. We can learn from their attitude to food and refection.[3]

The Okinawa Centenarian Study

The tiny Japanese island of Okinawa is notorious for its healthy elderly population. In 2001, when the Okinawa study was conducted, it had 427 people over the age of 100, enjoying robust good health. The study found that elderly Okinawans have among the lowest mortality rates in the world from a multitude of chronic diseases of aging and as a result enjoy not only what may be the world's longest life expectancy but the world's longest health expectancy. These centenarians have a history of aging slowly and delaying or sometimes escaping the chronic diseases of aging including brain degeneration with NCI, cardiovascular disease (coronary heart disease and stroke) and cancer. The Okinawa research group has investigated over 600 centenarians to uncover clues to their outstanding health and long lives. The goal of the Okinawa Centenarian Study was to uncover the genetic and lifestyle factors responsible for their remarkably successful aging phenomenon, for the betterment of the health and lives of all people. Studies have concentrated on the genetics, diet, exercise habits and the spiritual beliefs and practices of the Okinawan elders. The secret of the Okinawans was their attitude toward food and the manner of feeding. They have an almost religious reverence for the process of refection. The don't eat too much, they eat in a social setting and they take time and are calm, not rushed. When Okinawans move away from their island and adopt the usual modern Japanese foods and rushed style, they become susceptible to the same diseases as the average Japanese person. This means that their longevity and health is not all genetic. Their lifestyle has a lot to do with it. Calorie restriction, (hara hachi bu) is fundamental. (It takes the stomach twenty minutes to signal the brain that it's full. Stop eating when 80% full and wait 20 minutes after which time you will not be hungry.) They also avoid distractions while eating, which interfere with body's signals to stop. They wait for *twinges of fullness*. The lesson from the above is that we should eat at a leisurely pace and it should be a relaxed social event. We should not eat to be full and we should avoid animal protein and fat except for fish. That is the Okinawa diet – and of course green tea!

Is there an obesity epidemic?

Depending on what study one reads, estimates on obesity rates range widely. While some studies report that 60% of USA citizens are clinically obese requiring treatment, more conservative estimates are in the neighborhood of 30%.[4] About 25 years ago a survey by Harper's showed that people ranked their activities as TV watching, eating and shopping as the top three, in that order.[5] Are we not reaping the results?

Reputable studies show that age-adjusted prevalence of obesity increased by 30% between 1984 and 1994 (USA).[6] Obesity and its treatment consumes about 6% of the total USA health expenditure.[7] In Canada, obesity amongst children has increased dramatically and so has diabetes, a condition that was rare in children a generation ago. Combined childhood and adult obesity surveys show that as many as 60% are overweight and 30% are obese. Between 1981 and 1996 obesity prevalence for boys went from 2% to 10%; for girls it went from 2% to 9%. That certainly indicates a childhood obesity epidemic.[8]

There is little doubt that obesity is associated with increased morbidity and mortality. Obesity is associated with several serious ailments that are independent risk factors for early onset brain degeneration with NCI, Obesity itself is an independent risk factor for brain disease, particularly stroke. The higher the body mass index (BMI) the greater the risk of stroke. Obesity at age 40 doubles the risk (100% increase) of early onset NCI. In 1991, 15% of Americans were obese. By 1999, it was 27%. In 2002, 35% of Americans were obese. 10% of Americans under 17 were obese.[9] Obesity is becoming an epidemic of the poor also affecting people in developing countries.[10] In Western Europe and the USA, 0.5% of the population becomes obese each year. In China and Brazil, 1.0%. In Mexico and

S. Korea, 2.5% (ca. late 1990s). In India 33% of women are malnourished, 12% obese, even within the same household.[11]

Is your weight healthy for your height?

The Body Mass Index (BMI) is a measurement that tells us if our weight is in a healthy range for our height. (Explanation: BMI takes into consideration weight and height of a person and is expressed as kg/m^2 = kilograms per square meter). You can calculate your BMI using the chart below.

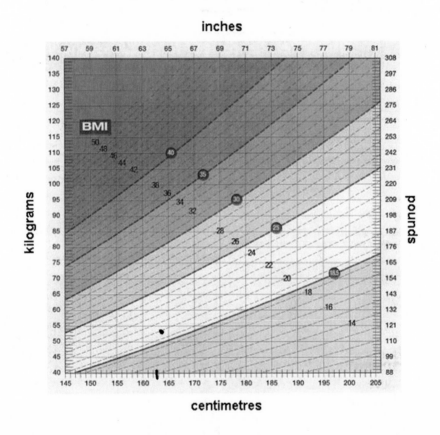

Once you have your BMI, use the following table to assess where you stand.

Health risk classification according to body mass index (BMI)

Classification	BMI category	Risk of developing health problems
Underweight	Less than 18.5	Increased
Normal weight	18.5 - 24.9	Least
Overweight	25.0 - 29.9	Increased
Obese class I	30.0 - 34.9	High
Obese class II	35.0 - 39.9	Very high
Obese class III	40.0 or more	Extremely high

For persons 65 years and older the 'normal' range may begin slightly above the BMI of 18.5 and extended into the 'overweight' range. To clarify the risk for each individual, other factors such as lifestyle habits, fitness level and presence or absence of other health risk conditions also need to be considered. For example, physical fitness in an overweight person will confer some protection.[12]

Check your body fat using the table below:

BMI categories	Percentage of body fat
Underweight	18.5 %
Normal weight	18.5-24.9 %
Overweight	25-29.9 %
Obesity	30 % or more

Abdominal obesity syndrome

Abdominal obesity syndrome carries the highest risk for complications of obesity. People who have high waist circumference and the resulting apple shape, have a higher risk for a vascular event, particularly if it's combined with a high BMI. Waist circumference is a better gauge of morbidity risk than the BMI.[13] The limit for North American men is 40 in. (102 cm) and for North American women it is 35 in. (88 cm). For Chinese men it is 34 in. (85cm) and for Chinese women it is 32 (80 cm). For Japanese men the limit is 34 in. (85 cm) and for Japanese women it is 30 in. (74 cm). As you can see there is quite a bit of ethnic variability.

Overnutrition syndrome

The *Overnutrition syndrome* is also known as *Syndrome-X,* and *Metabolic syndrome.* Untreated it may lead to the CHAOS syndrome. CHAOS is an apt acronym for when the metabolism goes haywire. It is characterized by Cardiac disease, Hypertension, Adult diabetes, Obesity and Stroke. Like obesity and the Overnutrition syndrome, CHAOS syndrome is lifestyle sensitive and it may be reversible in its early stages. Many of these people will need medical treatment. Sustained obesity is actually unnatural. If we look at the general population, as age progresses, the tendency is to eat less and weigh less. Caloric intake, protein intake and the consumption of fats gradually decreases; therefore, so does weight. That is natural, healthy aging. Don't forget that after the menopausal years, in both men and women, a state of negative protein balance prevails. This is determined genetically and hormonally and it cannot be altered by eating more protein. Essentially, after menopausal years, we lose a little more protein every day than we have gained. Therefore, the weight that is gained after the menopausal years is merely fat which will overload the cardiovascular system. It's good to have some body fat – we don't have to be skinny. In populations of the oldest-old, a bit of fat, but not obesity, appears to be associated with longevity.[14]

We certainly know that overweight and obesity are a result of too much caloric intake and not enough caloric expenditure. Given that one pound of body fat represents 3,500 calories, you can see that it is easier to regulate body fat by reducing intake than by exercising. Exercise is for fitness and fitness does influence metabolism – in a very good way. We will discuss the benefits of physical exercise in Chapter 9. For a sneak preview, think about the following:

1 donut = 1 beer = 1 soft drink = 200 kcal = walking briskly for 1 hour

We have talked about food and feeding quite a bit but we have not defined just what is food. What is good food? What is bad food?

What is food, anyway?

Moreover, what do our body and brain really need? The brain needs only glucose and oxygen from the blood to function properly, at least for short periods. If it does not get glucose or oxygen for more than a few minutes the brain actually suffers irreversible damage and dies. Much of what we eat is really to keep our body in good condition so that it can supply the oxygen and glucose and vital nutrients to the brain. We may say that in a person with adequate nutrition there is no single food that will directly improve the brain. Ironically, caloric restriction, short of producing malnutrition, remains the only known way to slow age related degeneration in higher organisms.[15] We suspect that this is the case for humans. Vegans do live longer – if they do not get protein, vitamin B_{12} and iron deficiency. It seems that the healthiest way to be is in a state of semi-starvation with an adequate supply of vital nutrients (minerals and vitamins). That kind of lifestyle is not likely to catch on voluntarily, because for most of us such a life would be too miserable. The Okinawa Study and others do suggest that caloric restriction has powerful effects against physiological changes associated with aging.[16]

What is food and what is drug?

Ancient, pre-Hippocratic medicine did not differentiate between what was food and what was medication. Ancient Chinese wisdom said "let food be your medicine and medicine be your food". Through the modern pharmaceutical industry we have gained access to very potent molecules that must be taken in very small amounts. These are substances we understand as drugs. We know that other substances found in herbs or animal products, in a more or less natural form, also have physiological effects on our bodies and brains. Some herbs have strong psychotropic, even hallucinogenic properties. The problem with herbal preparations is not that they don't do anything – it is just the opposite, they have efficacy. Unfortunately, it is difficult to harness their efficacy because herbal and animal products are difficult to standardize and regulate with respect to potency and purity. As a result, rigorous research on the effect of natural substances and even foods is lacking. The situation is not likely to improve because the boundaries between drugs, herbs and foods are truly blurred. Well-designed, prospective case controlled trials will not be undertaken on any herb or food that cannot be reliably manufactured and patented for profit. Therefore, we remain guinea pigs in a giant naturally occurring experiment, subjects to the scientists who study, retrospectively, the outcomes of various nutritional behaviors.

Take a look at the following list and see if you can decide whether the item is a food, natural medicine, herb, drug or a poison:

Alcohol – hard liquor	Alcohol – wine or beer	Chamomile tea
Chocolate	Coca-Cola	Cod liver oil
Cod liver	Coffee	Dried thyroid gland
Fish oils	Flax seed oil	Fructose
Fruit juice, fresh	Fruit juice with sugar	Fruit juice with caffeine
Garlic	Grapefruit	Grape seed oil
Ginger	Ginkgo	Ginseng
Glucose	Honey	Hormones of horses
Liver	Milk with vitamin D	Olive oil
Omega-3 oils	Onion	Potassium chloride
Potassium nitrate	Refined (white) flour	Refined (white) sugar
Sodium chloride	St. John's wort	Sucrose

Sweet calomel	Tea	Pig testicles
Tobacco	Vinegar	Water (H_2O)
Willow bark tea	Xylitol	Xanthine

The above is a partial list of substances we consume without much thought as to whether they are foods or drugs but they all have powerful, sometimes even deadly effects on body and brain. Some of them, like caffeine and alcohol, are highly psychoactive – acting primarily on the brain. Some of them are natural depending on how you define 'natural'.

The foxglove plant gave us digitalis, a commonly prescribed cardiac drug. Taking foxglove would have some effect but the dosing would be so difficult and inaccurate that the treatment would kill you long before the disease. I would much prefer pharmaceutical grade, synthesized digitoxin. Similarly, dried thyroid glands from pigs, beef or sheep yield thyroxin, a thyroid hormone that is prescribed for hypothyroidism, a very common disorder. Again, the guesswork of getting the right dose is dangerous – people have died from such ventures. I would prefer the 'unnatural' pharmaceutical grade manufactured thyroxin. I would feel unnaturally alive and well which is much better than being naturally psychotic from 'myxedema (hypothyroid) madness' that came with no treatment at all. The 'natural' dried thyroid gland treatment had a major side effect – very fast irregular hearbeat and death from heart attack. That was probably a better outcome than myxedema madness.

Food-Herb-Drug Interactions

Grapefruit, Valencia oranges, cranberry and some other fruits and juices inhibit liver enzymes that detoxify other substances. Watch out for drugs that may accumulate and become toxic when eating grapefruit, especially when starting for the first time, while on the following drugs : calcium channel blockers, cyclosporine, statins, benzodiazepines, carbamazepine, antihistamines, clomipramine, anti-parasitics, haloperidol, sertraline, ethinyl estradiol, cisapride, and more. That is a partial list. I included it just to prove a point – foods, drugs and herbs can have powerful interactions. There are many other examples but my purpose is not to turn you into a pharmacologist. Let you doctor be the guide. You may also access drug interactions on the Internet.

It is mankind's nature to yearn for what is natural and even mysterious. As long as there will be plants and animals to sample, people will experiment. More regrettably, there are many cases of horrible diseases which cannot be cured by medical science; therefore, victims will develop false hopes and even turn to magic. When using herbs as medicine, please keep the following in mind:

Health Hazards: herb & food controversies

- Encourages self-medication without diagnosis
- Expensive; supplies are not always available
- Few randomized, controlled studies
- Illusion of natural safety and control over health
- Poor quality control; labeling is not uniform
- Unregulated industry; classified as food
- Variable concentrations; dosing is guesswork

Some herbals and home remedies have a salicylate effect (reduced blood clotting tendency). Those include garlic, gingko, ginseng, ginger, antlers and rhino horn, to name the common ones. Taking several of the above, particularly when combined with aspirin (salicylate) or other anticoagulants like warfarin, has been known to cause cerebral haemorrhage.

An antidepressant (MOA-inhibitor), although highly efficacious, causes a spike in blood pressure when combined with tyramine containing foods, which include, pickles, cheese, red wine, avocado or cough syrup. The sudden rise in blood pressure has caused strokes and death; therefore, doctors avoid prescribing this drug unless there is no alternative. Scary stuff, this world of drugs, foods and herbs!

Commonly consumed substances that merit special attention

Water

We don't think of water as food or nutrient but without it nutrients cannot dissolve and enter the body. In chemistry water is known as "the universal solvent". Water is a chemical compound and its concentration is very tightly regulated in the blood. Too little or too much will cause delirium and death. Water is the principal solvent in the body and all chemical reactions of life take place in an aqueous medium. The lean body mass is about 70% water followed by proteins, fats, minerals and carbohydrates. Water is not a nutrient, it has no energy (caloric) value. In fact it takes about 10 calories to raise the temperature of a 12 ounce glass of cold water to body temperature; therefore, there is actually a very small net caloric loss after drinking water. I find that most people, particularly older people, do not drink enough water. Most of my older patients are actually dehydrated. This can be very dangerous, especially when a person is on a number of medications because in a state of relative dehydration, blood concentrations of drugs increase and may reach toxic levels. A person is most dehydrated in the morning, after a night of fasting, so the best time to load up on water is in the morning. Drinking a glass of water before a meal has important beneficial effects. It provides some stomach distention and a feeling of fullness, which will naturally suppress appetite and tend to prevent overeating. It keeps the food moist and interacts with fiber to provide bulk and lubrication that is essential for proper digestion and bowel motility. Water balance must be maintained so that the amount gained equals the amount lost, otherwise water intoxication or its opposite, dehydration will result. Typically, for the average person, it looks like this:

Average daily water intake:	Average daily water loss:
Liquid food (water, soup, milk, etc) 1100 ml	Vaporization 920-1000 m
Solid food (moisture) 500-900 ml	Defecation 80-100 ml
Water from oxidation (metabolism) 400 ml	Urination 1000-1300 ml
Total intake: 2000-2400 ml (=2.4 L)	**Total loss:** 2000-2400 ml (=2.4 L)

Vaporization includes sweating and moist air from the lungs. There is also some caloric loss because it takes about a half a calorie to vaporize 1 ml (1 gm) of water. So you lose a few calories from drinking cold water and also from vaporizing it – a good deal for those of us who are battling the bulge.

It is a good habit to reduce or stop water intake in the evening in order to avoid frequent trips to the bathroom. Fragmented sleep is a form of sleep deprivation that carries a risk of morbidity. More about sleep in Chapter 11.

Coffee

Surprise! Coffee may be good for you. A study, which followed 27,000 older women for 15 years, found up to a 30% reduction in the risk of cardiovascular disease in those people who had a moderate intake of coffee.[17] One to three cups a day may protect people from heart disease and strokes, according to new research contradicting numerous studies that have suggested that coffee is bad for you. The analysis, part of the Iowa Women's Health Study, found that up to 60% of antioxidants in the diet may come from coffee.[18] Antioxidants protect cells from damage and reduce the inflammation that encourages arteries to narrow. Active ingredients of coffee include caffeine and polyphenols. Polyphenols are also found in red wine and they too have been linked to a reduction in the risk of cardiovascular diseases in people who drink one or two glasses of red wine a day. The researchers in the Iowa study also pointed out that a Scottish survey of 11,000 men and women found that coffee drinking was associated with a reduction in deaths from all causes. Dr. Sarah Jarvis, a fellow of the Royal College of General Practitioners said: "This is a message about moderation. Too much exercise, too much coffee or too much alcohol are bad. In moderation they are beneficial".[19] Unfortunately, caffeine is a stimulant that causes vasoconstriction, a narrowing of the small arteries and that contributes to hypertension, a risk factor for coronary heart disease and stroke. So who do you believe? The above is a good example of a common substance, that may benefit some but harm others. If your blood pressure is normal, coffee is no problem. Coffee also appears to reduce the risk for diabetes. The merit of coffee remains a complex issue but despite its association with poor life style habits, heavy coffee consumption appears to be protective for diabetes. A recent study from Harvard School of Public Health showed that people who drink a cup of coffee six or more times a day may reduce their risk of diabetes by more than 50%.[20] During the Nurse's Health Study cohort involving 84,276 women from 1980 to 1998, coffee consumption was documented every 2-4 years through validated questionnaires. The researchers found an inverse relationship; the more coffee consumed the lower the incidence of diabetes.[21] The Health Professional Study of 41,934 men, aged 40-75 were followed with food frequency surveys and asked to report diabetes diagnosis between 1986 and 1998. Results were similar to that of the Nurse's Health Study.[22] Other studies have also indicated that coffee's blood pressure elevating effects are short lived. It's difficult to ascertain how coffee exercises its effects because there are hundreds of compounds in the coffee bean – including opioids. It may be that potassium, niacin, magnesium and anti-oxidants such as tocopherols, phenols, cholinergic compounds and opiate receptor stimulants in the coffee are protective.

The full chemical name for caffeine is 1,3,7-trimethylxanthine and its chemical formula is $C_8H_{10}N_4O_2$. Caffeine was first isolated from coffee in 1820. In its pure state, caffeine is a crystalline white powder. Caffeine can be found in 60 different plants. About 10 grams of caffeine is considered a lethal dose (A cup of regular coffee contains about 50 milligrams = 50 thousandths of a gram.) Caffeine is the most popular drug in the world. It is added to many soft-drinks and even some beers! Ninety percent of Americans consume it in some form every day. Darkly roasted coffee has less caffeine than lightly roasted. Caffeine can cause insomnia; therefore, it's best to have it in the morning hours. *Bon matin!*

Tea

Tea and its compounds are attracting new research attention. Tea's popularity has grown in the last 10 years; it is the world's second most favored drink, after water. In the USA, more than 37 billion glasses of iced tea are consumed yearly. One cup of tea contains nearly half of the caffeine (about 25 mg) found in a cup of coffee and no calories if taken straight. Its reputation as being protective against some diseases is growing. Teas are a rich source of the flavanoids, polyphenols and catechins – all powerful antioxidants that protect the body's cells and tissues against damage from free radicals. A single cup of tea contains more antioxidants than one apple. Research suggests that tea's powerful antioxidants can lower cholesterol and blood pressure, and even help in the recovery from heart attack. In recent years some of the epidemiological studies have shown that drinking green or black tea may be beneficial in preventing some cancers.[23] According to a USA study, five servings of black tea daily lowered LDL cholesterol by 11.1% and total cholesterol by 6.5% in people with mild elevations.[24] Tea has been noted to prevent or benefit hypertension which is an independent risk factor for cerebrovascular disease and NCI.[25] For quite some time researchers have suggested that drinking regular cups of tea could help improve your memory. Now we have proof that green tea enhances connectivity during memory processing – we can actually see it on functional magnetic resonance imaging (fMRI).[26] A team from Newcastle University found green and black tea inhibited the activity of key enzymes in the brain associated with memory loss. The researchers hope their findings may lead to the development of a new treatment for AD. They say tea appears to have the same effect as drugs specifically designed to combat the condition. AD is associated with a reduced level of a neurotransmitter chemical called acetylcholine in the brain. In lab tests, the Newcastle team found that both green and black tea inhibited the activity of the enzyme acetylcholinesterase (AChE), which breaks down that key chemical. They also found both teas inhibited the activity of a second enzyme butyrylcholinesterase (BuChE), which also breaks down acetylcholine. Green tea went one step further in that it obstructed the activity of beta-secretase, essential in the production of the harmful protein deposits in the brain associated with AD. The scientists also found that it continued to have its inhibitive effect for a week, whereas black tea's enzyme-inhibiting properties lasted for only one day. Drugs like rivastigmine, galantamine and donepezil, marketed for AD work in a similar way. The Newcastle researchers plan to carry out further tests on green tea and they hope they will include clinical trials. Their aim is to work toward the development of a medicinal tea specifically aimed at Alzheimer's sufferers. The next step is to find out exactly which components of green tea inhibit the activity of the enzymes AChE, BuChE and beta-secretase. Lead researcher Dr. Ed Okello said, "Although there is no cure for Alzheimer's, tea could potentially be another weapon in the armory which is used to treat this disease and slow down its development. It would be wonderful if our work could help improve the quality of life for millions of sufferers and their caregivers. Our findings are particularly exciting as tea is already a very popular drink, it is inexpensive, and there do not seem to be any adverse side effects when it is consumed. Still, we expect it will be several years until we are able to produce anything marketable".[27]

Black tea, traditional English breakfast tea, is derived from the same plant as green tea, *Camellia sinensis*, but has a different taste and appearance because it is fermented. A Taiwanese study found a correlation between steady long term tea consumption and low body fat but it is not known whether tea really causes fat loss.[28]

Cocoa and chocolate

Chocolate is an ancient food made from roasted beans from the *Theobroma cacao* tree. The Aztecs supposedly had the word *chocolatl* which worked its way into first Spanish, then English. Ancient Malayans referred to it as the *Food of the Gods* because of its very interesting nutritional and psychotropic effects. There is no doubt chocolate has psychotropic effects and perhaps for that very reason, it is striving to find a place in modern medicine. Cacao beans are also rich in flavonoids, which are powerful antioxidants. (Antioxidants will be discussed in detail later in this chapter.)

Food or drug? Although addictive behaviour is generally associated with drug and alcohol abuse or compulsive sexual activity, chocolate may evoke similar psychopharmacological and behavioral reactions in susceptible persons. A review of the literature on chocolate cravings indicates that the pleasurable appeal of chocolate (fat, sugar, texture and aroma) is likely to be a predominant factor in such cravings. Other characteristics of chocolate may be equally important contributors to the phenomena of chocolate cravings. Chocolate may be used by some as a form of self-medication for dietary deficiencies (e.g. magnesium) or to balance low levels of neurotransmitters involved in the regulation of mood, food intake and compulsive behaviours (e.g. serotonin and dopamine). Chocolate cravings are often episodic and fluctuate with hormonal changes just before and during the menstrual cycle, suggesting a hormonal link and confirming the assumed gender-specific nature of chocolate cravings.[29] Chocolate contains several biologically active constituents (methylxanthines, biogenic amines and cannabinoid-like fatty acids and powerful antioxidants), all of which can potentially affect mood and behaviour. Thus, chocolate cravings, mostly in women, are real. The psychopharmacologic effects of chocolate must be considered when formulating recommendations for overall healthful eating and for treatment of nutritionally related health issues. Could chocolate treat mood disorder in some people?

Dark chocolate, with the highest cocoa content, has the highest concentration of flavonoids. Most North Americans, however, indulge in milk chocolate where the extensive processing has stripped away much of the natural beneficial compounds. Although chocolate does contain saturated fat, one third of the total fat comes from stearic acid, not known to raise low density lipoproteins (LDL – the 'bad cholesterol'. Dark chocolate contains 70% cocoa butter, a source of stearic acid, while commercial candy bars contain only 20%. Contrary to popular belief, chocolate is only a weak stimulant. In an Italian study, participants who ate 3.5 ounces of dark chocolate daily for 15 days had a 12 mmHg decrease in systolic blood pressure and a 9 mmHg drop in diastolic blood pressure compared to those who ate white chocolate.[30] Scientists have found that theobromine, a main ingredient in chocolate, is more effective in stopping persistent cough than codeine medications.[31] The Kuna Indians of Panama, who drink flavanol-rich cocoa made from locally grown beans, have no hypertension. High blood pressure develops in those who move to the mainland and start drinking commercial cocoa.[32] Cocoa flavanols have a number of beneficial effects: improved heart health, better artery health, lower blood pressure, lower cholesterol and diabetes prevention. High heat and commercial processes to extract the 'chocolate' from cocoa tend to destroy the flavanols. Interestingly, it is a candy company (Mars Inc.) that is experimenting with better extraction (patented) of flavanols and is preparing a 'chocolate pill' that will be extremely rich in flavanols. The hope is to use that in the prevention and treatment of diabetes and cardiovascular disease. My hope is that it will have similar benefits for cerebrovascular disease. The researcher, Dr. JoAnn Manson, Head of Preventive Medicine at Harvard affiliated Brigham and Women's Hospital, in Boston, is a leader in the field.[33]

Even though an ordinary chocolate bar does exhibit strong antioxidant activity, the health benefits are dubious because of the relatively large amounts of saturated fats (8 g per 40 g bar) and sugar that are added for marketability. Dark (85%) chocolate is better but not as sweet. Pure chocolate is actually bitter – not at all sweet. However, a cup of hot cocoa has a much lower level of saturated fats (0.3 g per serving). As noted previously, research shows that cocoa has high flavonoid content and substantial antioxidant capacity. On a per serving basis, 4-5 times stronger than black tea, 2-3 times stronger than green tea and almost twice as strong as red wine.

Alcohol

Chemically and pharmacologically, alcohol is classified as a central nervous system depressant – an anaesthetic. Studies show that about five out of six Caucasians of European ancestry seem to benefit from moderate alcohol consumption. They can certainly tolerate higher amounts than the other people.[34] What about the rest of the world? There are great variations in alcohol dehydrogenase and other detoxifying liver enzymes (genetic polymorphism). Alcohol is addicting. The drinker will need more and more to get the same effect. Many people find alcohol highly toxic. It was always believed, and it may be true, that alcoholic beverages have a number of beneficial effects. In moderation, it is a relaxant, improves circulation and protects against vascular diseases like coronary artery disease and cerebrovascular disease.[35] Recently a study showed that it offered some protection to those with multiple sclerosis.[36] In particular, red wine is believed to be protective because in addition to the above, it contains the powerful antioxidant resveratrol. Some studies have shown a protective effect, particularly for red wine, against mortality from all causes, coronary heart disease and cancer. An often quoted study called the *Copenhagen Study* examined the type of alcohol consumed and mortality. The study looked at 13,064 men and 11,459 women between the ages of 20 and 94. This represented 2,578,859 person-years of follow up. The results were adjusted for age, gender, education, smoking, physical activity and BMI. Light drinkers, especially wine drinkers, who had 7-21 drinks per week appeared to benefit. Heavier consumption of alcohol in any form was associated with increased mortality.[37]

"Put down that glass of red wine!" a medical reporter wrote recently. Contrary to what many earlier studies have shown, a research project from the University of Victoria suggested that alcohol may not protect against heart disease after all. "Apologies. It must be so confusing," said Dr. Tim Stockwell, director of the University of Victoria-based Centre for Addictions Research of British Columbia. His group has found that most studies that maintain moderate drinking helps prevent heart disease are flawed. It's quite probable that drinking alcohol offers health benefits, "...but maybe they are smaller than have been estimated," he said. "Just don't stake your life on it. Just drink very cautiously. At low levels it is not harmful, it may even give you a little benefit." The international team was led by researchers from University of Victoria and the University of California. It analyzed 54 studies conducted from 1974 to 2004 linking how much people drink with risk of premature death from all causes, including heart disease. Researchers investigated an earlier claim that many studies about the link between moderate drinking and heart disease made a critical error. Those studies included as abstainers people who had reduced or quit drinking because of chronic illnesses, declining health, frailty, drug use or disability. When abstainers were compared with moderate drinkers, the imbibers came out slightly healthier. However, seven studies that compared moderate drinkers to long-term abstainers found no difference in risk of death.[38]

Despite all the folklore and rationalizations for the alcohol habit, the cost of alcohol-related brain problems remains huge. Annual cost of hangovers in the USA through absenteeism and poor job performance is around $US 148 billion (2007). Binge drinking, the most harmful style, has increased in Europe and USA. We will not even try to attempt to quantify the damages from violence, accidents and the cost of brain damage caused by alcoholism. Recent reports indicate that governments are worried; 30% of fatal accidents in Europe are linked with alcohol. When illness, lower output and other social costs are taken into account, alcoholism costs Europe and America hundreds of billions of dollars per year – up to 1.5% of GDP! No wonder governments are trying to increase costs through taxation to reduce consumption by destructive drinkers.[39] Cheers!

Bad nutrition

Foods can have some very powerful effects. Even when they have a weak effect, multiplying this over a period of time, usually many years, may have profound effects on our bodies and brains. What happens inside the brain, of course, affects cognition, mood and behaviour. Should we not take foods more seriously? Let us look at what is *good nutrition* and *bad nutrition*. We often use the term malnutrition to mean poor nutrition and that evokes images of people who are undernourished and wasting away. We generally do not think of people who are overweight or obese as having malnutrition. In fact, the term malnutrition means 'bad nutrition' and under-nutrition and over-nutrition are bad for health. Clearly, not enough food intake will cause weight loss and this is likely to be accompanied by deficiency states. A high caloric intake will cause weight gain but does not guarantee the absorption of essential nutrients. Essential nutrients include vitamins, minerals, essential amino acids and essential fatty acids. These substances cannot be made by the body and have to come from our dietary intake. Vitamin D is an exception to this because it is not likely to come from food sources. It is actually a hormone that is made by the skin and acts upon various parts of the body, including the brain. We will cover this interesting substance in the chapter on hormones (Chapter 13).

Good nutrition is characterized by a high nutrient-caloric ratio. I like Dr. Fuhrman's definition of nutritional health as H = N/C. (Health value = nutrients/calories).[40] An example of a food with a high N/C is broccoli. An example of a food with a very low N/C is popcorn. Popcorn is not bad for you it just happens to be mostly air and starch. Popcorn would be worse if it had no air in it. Potato chips have an even lower N/C. Foods that have a very low N/C are often referred to as foods that have *empty calories*. This phrase refers to the fact that such foods have high caloric value but carry no nutrients; therefore, they will cause weight gain without supplying nutrition. People who eat a lot of empty calorie foods can become obese and develop vitamin and other nutrient deficiencies at the same time. Although the brain uses 25% of our caloric demands it also needs nutrients. Unfortunately, the modern American diet (MAD) is deficient in nutrients. Fast foods, processed foods, high fat foods, starchy foods and generally all those delicious sweet, salty and fatty snacks have a very low N/C. The MAD nutritional lifestyle will kill a person but it will take a number of years. That is not good for your health span, your brain health or your life span.

You don't need to turn into a nutritionist; simply understanding the basics is a good start. It is impossible to talk about nutrition for the brain without having a bit of knowledge about the nutrition of the whole body.

The indirect and direct effects of bad nutrition on the brain

Bad nutrition can have *indirect* and *direct* effects on the brain. *Indirect effects* include dietary habits that cause a condition that increases the risk of brain disease and NCI – usually cerebrovascular disease or stroke. For example, hypertension can be caused by high salt intake and that can cause vascular disease in the brain or stroke. A study showed that a low salt diet reduced blood pressure in those with normal blood pressure by an average of 7.1 mmHg and in those with high blood pressure by 11.5 mmHg. That was a randomized prospective study for 30 days involving 412 people. The researchers concluded that long-lasting dietary changes are most important.[41] Another example of an indirect effect is seen in the development of obesity. Gradual development of overweight and obesity, especially when coupled with physical inactivity, leads to diabetes and, in turn, leads to vascular disease and progressive NCI. Indirect deleterious effects on the brain, as a result of chronic malnutrition, are actually quite common. Can you think of some more examples of indirect effects of poor nutrition on the brain? What about a high cholesterol (animal fat) diet, even without weight gain? Did you know that more than one soft drink per day, whether juice or diet, is associated with obesity, heart attack and stroke?

Direct effects are simpler in their mechanism. A direct effect on the brain, from improper nutrition, is either a deficiency of a vital nutrient or the ingestion of a brain poison (neurotoxin). Let us look at deficiencies first: Deficiency syndromes have a characteristic presentation but these are so subtle at first that they are easily missed or misdiagnosed. What we tend to see clinically are the late stage manifestations. For example vitamin B_{12} deficiency can cause progressive NCI and irreversible brain damage. Folic acid deficiency, historically seen in young women, can cause *spina bifida* in their babies. (This condition is prevented because the government now adds folic acid to cereals). That is not exactly a brain disorder but is very closely related to brain development and has cerebral developmental consequences. People with folic acid deficiency in later years develop a brain syndrome which involves cognitive and mood changes. The following table shows some of these relationships:

Deficiency/Group	Symptoms
Folic acid (Low vegetable diet)	Poor memory, depression
Vitamin B12 (Low meat diet)	Progressive NCI, Dementia, numbness, tingling of extremeties, poor balance
B vitamins (Vegetarians, low meat diet)	Brain syndrome, progressive NCI,

Iron (High starch, low meat, low vegetable diet)	Anemia and decline in cognitive and physical performance
Iodine (Low salt, low seafood intake)	Hypothyroidism, NCI, depression
Vitamin D (Lack of sunshine, low milk and liver oil intake, no supplementation)	Depression, low energy, muscle weakness
Vit C (Lack of fruits, vegetables, smoking)	High capillary fragility, cerebrovascular disease

Some deficiency states are common; often they are restricted to a cultural or geographic area. A good example is iron deficiency. Iron deficiency can impair young women's memory and attention span but iron supplements can help. The research was done on women because they are known to be at high risk for anemia. The women were divided into three groups: normal; iron deficient but not anemic; or anemic (a severe iron deficiency with decreased oxygen-carrying red blood cells). About half had mild iron deficiency without anemia. At the beginning of the study, the women performed a set of tests to measure attention span, memory and learning. On the first test, women who were iron deficient but not anemic completed the tests as quickly as those with normal iron levels but they made more mistakes. After the first round of tests, the women were assigned to take a daily iron supplement or a placebo for four months. The remaining women were then retested. Women who took supplements and regained healthy iron levels improved their test scores and speed (after taking into account differences such as smoking and grade point average). The study shows the significance of iron deficiency on cognitive skills and suggests that supplements help performance.[42] Advice: Consume foods high in Vitamin C, which enhances iron absorption. Cook with iron pots and pans. Drink a glass of orange juice with your meal. Eat foods rich in iron including oysters, liver and other organ meats, lean red meat, tuna, eggs, dried fruit, beans and dark, leafy vegetables. The researchers said, in industrialized countries about 10% of women of reproductive age and 25% of pregnant women are iron deficient. In non-industrialized countries, the prevalence of anemia is over 40% in non-pregnant women. Consequences include fatigue, reduced physical endurance, impaired immune response and temperature regulation difficulties. The re-introduction of cast iron pots and pans in some regions of Africa has improved endemic iron deficiency.[43] Scientists are starting to recognize the health effects of iron deficiency before it reaches the level of anemia. Doctors advise women to have their iron levels checked before buying iron supplements. Extra iron doesn't help those with normal levels and too much iron can lead to liver damage.

The above are just some common examples of deficiency states. There are a few more uncommon and esoteric syndromes but they are hardly ever diagnosed or documented. How many people have you known who suffered from decreased mental alertness as a result of boron deficiency? The syndrome does exist but have you even heard of such a thing? As you can see there is some overlap in the classification. It is difficult to keep direct and indirect effects separate. Iodine deficiency, for example, causes hypothyroidism which can parade as NCI and depression. That could be thought of as an indirect effect because the brain itself doesn't use iodine directly. Similarly, vitamin C deficiency, causing capillary fragility, affects the vascular system all over the body. The effect of vitamin C in the brain remains a bit of a mystery. We know that vitamin C (ascorbic acid) crosses the blood-brain barrier and it is found in the brain tissue in much higher concentrations than in other organs. Some researchers have also noted that elderly people whose diets are rich in vitamin C are less likely to die from strokes. The latter research was conducted by Dr. Christopher Martyn, and it was reported in the *British Medical Journal* in 1995 as a retrospective study involving 643 people.[44] His research group found that low vitamin C intake was equal to high blood pressure as a risk factor for strokes in the elderly. Vitamin C intake in the top third had less than half the risk of dying from stroke compared to lower third.

Vitamin C, also called ascorbic acid, first isolated in 1928, was identified as a cure for scurvy in 1932. Vitamin C is a water-soluble, carbohydrate-like substance that is involved in certain metabolic processes in animals. This vitamin is essential in a variety of metabolic functions, including synthesis of collagen (for connective tissue), maintenance of the strength of blood vessels, metabolism of certain amino acids, and the synthesis and release of hormones in the adrenal glands. Fairly large amounts are needed by mammals. Humans, some monkeys and the guinea pig cannot synthesize ascorbic acid and must obtain it through food. Citrus fruits and fresh vegetables are the best dietary sources of ascorbic acid. Raw meat and raw fish can also provide vitamin C. Vitamin C is not stable under high heat conditions and is destroyed by cooking.

There has been a resurgence of interest in vitamin C. Linus Pauling, the American biochemist and two-time Nobel laureate (deceased 1994) was convinced that vitamin C played an extremely important role in vascular health. The German physician Matthias Rath came to the conclusion years ago that atherosclerosis, otherwise known as hardening of the arteries, was not a disease but a way for the body to repair or strengthen damaged blood vessels.[45] The idea is that it is not high cholesterol from diet that causes the problem but rather the lack of vitamin C. Vitamin C stimulates production of the protein collagen, which keeps arteries flexible. According to the new theory, when the body is unable to produce enough collagen to repair the arteries, it uses other lipoproteins, calcium and cholesterol instead, which lead to hardening and disintegration of the arteries. Lipoproteins are sticky molecules that transport fat and cholesterol through the body. The theory is substantiated by a study done by the National Heart Lung and Blood Institute. Studies showed that people with the highest intake of vitamin C had the lowest incidence of heart disease.[46] More recent research involving 300,000 Americans shows that taking at least 700 mg of vitamin C a day reduces the risk of heart (artery) disease by 25%. How much vitamin C we really need is controversial. A 70 kilogram (154 lb.) goat produces 13,000 mg (13 g) of vitamin C a day. A large dog produces 2.5 g a day. The American Committee on Animal Nutrition has also shown that monkeys, who like people are incapable of producing their own vitamin C, need a daily dose of around 55 mg per kilogram of body weight. That is what the veterinarian would recommend. Translated into a 70 kilo human that would amount to some 4,000 mg (4 g) a day, a dosage never recommended by medical doctors. Although recent studies support the old studies,[47] clearly more research needs to be done into the role of vitamin C and preventing and possibly improving vascular health – highly important in the brain and heart.

Studies show that various vitamin compounds with antioxidant properties can slow the progression of moderate AD. Vitamin C is a scavenger of free radicals, substances that play a role in causing the disease. Ascorbic acid must be converted to dehydro-ascorbic acid before it can penetrate the blood-brain barrier where it is converted back to ascorbic acid. Experiments are underway to investigate the use of dehydro-ascorbic acid as a way of getting higher levels of vitamin C into the brain.[48]

Neurotoxins

There are certain substances we ingest, willingly or unknowingly, that directly and rapidly affect the brain. The ingestion of neurotoxic substances is the ultimate form of malnutrition. We might use the oxymoron 'suicidal nutrition'. The substances are specifically toxic to brain cells and wreak havoc even in very small amounts.

Note that the caffeine and nicotine are not on the list. There is no evidence that caffeine is neurotoxic. Nicotine itself is not the main culprit in cigarettes and it is not why smoking is banned. Smoking cigarettes causes lung cancer and chronic obstructive lung disease independently of the nicotine. While nicotine is a strong stimulant to the brain and children have died of nicotine poisoning (convulsions and death), it cannot be taken through cigarettes, cigars or chewing tobacco in doses large enough to cause brain damage.

Alcohol

There is no doubt that excessive use of alcohol causes brain damage. There are so many books and journals on this subject that I will not sacrifice space here. As stated earlier in this chapter, governments are recognizing the costs, both in human misery and health care costs, and they are trying to limit the alcohol industry which is hell bent on increasing sales.

Artificial food colourings (AFCs)

These substances appear to be well tolerated by most people, however some children with ADD and ADHD are adversely affected by AFCs. Tartrazine, a common yellow food dye, is known to lower seizure threshold, at least in mice. For these reasons, some infant medicines come in dye and dye free versions. Some food colorings however, coming from unregulated sources from under-developed countries, contain lead compounds added to food as a colorant. We will pay more attention to this topic when we come to the section on lead poisoning.

Commonly abused drugs

It is not the purpose of this book to explore various drugs of abuse. That is a whole field in itself and there are many textbooks on the subject. It is tempting to conceptualize drug addictions (which are real diseases) using the *'infectious disease model'*. There are many similarities: The disease has a source and a reservoir, spreads from person to person, rips through large populations in cities where anonymity prevails and once epidemic proportions are reached, the disease is very difficult if not impossible to eradicate. The social and emotional cost becomes horrific – that is where we are today with respect to drug abuse. Drugs that damage the brain are a curse on brain health and people who promote them are criminals. Drug-induced brain damage is the major cause of unhealthy brains in young adults. Emergency rooms in large cities are flooded by younger adults having brain syndromes directly attributed to drug abuse, to the extent that they are actually preventing other people from getting health care. It's not a problem amongst older people simply because many of these drugs are so dangerous that people who use them do not get old – they stop using or die. The following is a partial list, not in the order of importance but in alphabetical order, of the most common drugs of abuse:

- Alcohol
- Amphetamines, 'crystal meth' and 'ecstasy' (related compounds)
- Cocaine
- Ketamine
- LSD and other hallucinogens
- Marijuana (cannabis)
- Opiates, morphine, heroin (related compounds)
- Volatile compounds (gasoline, solvents, etc.)

Cannabis, like alcohol, does not appear to be highly dangerous in small amounts but it is definitely associated with amotivational syndrome, depression and even psychosis.[49] It's been around for thousands of years but that does not mean that it is good. People seem to enjoy it because of its relaxant and stupefying effects, which is why it is called 'dope'. The medical profession and researchers have not found great uses for this drug although synthetic and purified forms exist for pain and nausea during palliative care. Some people with Parkinson's disease experience relief from tremor but this has not been well researched. Some strains of cannabis have anticonvulsant effects and research is going on to purify it for the treatment of a type of epilepsy.[49.1] Cannabis remains a drug looking for a disease.

Having said all that, it is important to acknowledge that like many other compounds and 'drugs', THC has been an *'orphan drug'* meaning that it has been ignored by the pharmaceutical industry because there is little profit in it – it cannot be patented and sold at high prices. To conduct a double-blind, prospective study with enough power to yield a meaningful scientific conclusion costs tens of millions of dollars. Who will do that for various extracts of marijuana? Marijuana, like tobacco, has hundreds of compounds in it and plants can be bred to exhibit particular properties or once identified, particular compounds could be isolated or synthesized. We know that there are compounds in marijuana that act as relaxants, anticonvulsants, antiemetics and analgesics – and maybe more! Too bad we do not do more research to access those effects. Legalizing marijuana for entertainment is not a step in the direction of medical utility. However, if it becomes legal, some people may undertake further legitimate research.

'Street drugs' are all associated with major psychiatric syndromes that are incompatible with autonomous functioning and often even with survival. To call them mind altering is euphemistic and optimistic; they are brain altering – and not for the better. I will have a few more words about this topic when I discuss treatment of risk factors for brain disease (Chapter 14).

Lead

Lead poisoning can be acute or chronic. I suppose the most acute form lead poisoning is a bullet in the head. In fact, lead toxicity itself is associated with violence. Lead poisoning is still with us and the situation varies from country to country. Children remain at the highest risk. Risks, whether from paint, drinking water or other environmental sources, are also high for women who are pregnant or nursing. Lead poisoning was recognized as a health threat decades ago, and it remains an issue today. The Centers for Disease Control (CDC) estimates that approximately 434,000 children in the USA already have lead levels greater than 10 µg/dL of blood, the agency's threshold for concern.[50] We do not have corresponding numbers for Canada, but we can assume that it is about one tenth of the prevalence south of the border (estimate about 50,000 children).

The CDC recommends testing children for lead exposure at ages one or two, and the American Academy of Pediatrics also urges physicians to provide advice to parents that could prevent lead exposure in the first place. For example, physicians could caution the many families who embark on renovation projects of older homes, when a new child is expected, that they run the risk of exposing a vulnerable fetus and any other young child in the home to lead from paint chips and dust. The CDC established its current lead exposure standard of 10 μg/dL of blood in 1991, and many believe that the level should be lower still. Telling parents whose children have lower than the 'accepted' lead levels that "all is well" is overly simplistic.

Other risks include dangers of herbal remedies leading to lead poisoning in newborns. One example: Severe lead poisoning in a pre-term infant was attributed to the mother's ingestion of an herbal remedy. The mother had gastrointestinal symptoms at 30 weeks gestation for which she was prescribed an herbal preparation containing lead and mercury. Her symptoms worsened (severe anemia, stomach symptoms, seizures). The infant had very high (25 times public health limit) levels of lead in the blood, was treated and discharged a month later with deafness and a two-month delay in neurological development. Lesson: Herbals are unregulated and often contaminated (from India, Mexico and Asia) with lead and mercury. Calcium supplements do contain lead but actually reduce absorption from all other sources which are 10 times greater. The net effect of calcium intake is to reduce lead levels. Vitamin C also reduces lead levels. A placebo-controlled study of 75 men, ranging in age from 20-35 years, showed that those who took 1,000 mg of vitamin C daily had dramatically reduced blood lead levels.[51]

Mercury

Nerve and brain cells are highly sensitive to mercury. Women who may become pregnant or who are pregnant should avoid large fish and canned fish because mercury may have had a chance to accumulate in larger, older fish from contaminated waters. Otherwise, mercury is not readily found in foods but has been (and in some countries it still is) a component of some medicinal compounds made to produce tranquilization and fever control. Mercury is also found as a stabilizer and antifungal treatment in seed grains which are for planting, not consumption. The scare about amalgam fillings appears to be largely unfounded. Studies show that people with amalgam fillings do not have a higher rate of brain-related problems than the general population. Having all old fillings removed is an overreaction but given a choice, it may still be wiser to opt for non-amalgam fillings in the first place. It is really a public health issue. Many more lives are improved and even saved by having good teeth and healthy gums, even with amalgam fillings (that are inexpensive and easy to perform), than without such dental care. In any case, mercury is highly toxic and we know that at least in animal models it produces lesions similar to those found in AD.

Nitrates

Potassium and sodium nitrate are added to many foods to preserve color and to prevent decomposition. Nitrates are found in high concentrations in meats, especially in, cold cuts, salamis, sausages and hotdogs. Nitrates are in so many foods it's hard to list them all. Even wine contains nitrates to give it a long shelf life under a variety of harsh conditions. Some people seem to tolerate high levels of nitrates but some are highly sensitive. It has been estimated that about 40% of people with tinnitus (ringing in the ears), which is a very commone ailment, are suffering from nitrate toxicity.[52]

Persistent organic pollutants (POPs)

Since the Love Canal disaster in 1978, POPs are no longer making headlines, but they are still lurking in the environment. The problem with these substances is that they persist and work their way into the food chain and into our tissues. The longer a complex organism (like a human being) lives, the more of these substances are accumulated. POPs have been identified in farmed salmon and other large fish. The bigger and older the fish, the more POPs. These substances have endocrine properties because they are similar to molecules found in hormones, particularly estrogens. We know that the brain is exquisitely sensitive to hormones (see Chapter 13) and there are grave concerns that these substances are causing endocrine disruption that becomes manifest as aberrant sexual development, reproductive and behavioural problems and cancers. There is little doubt that these substances affect brain development. Although most of the attention has focused on the sensational – the genital malformations and cancers – neuropsychiatric effects have been well documented. Prenatal or early postnatal exposure to POPs have been associated with learning and behavioural abnormalities. In *utero* and lactational exposure of nonhuman primates to environmentally relevant levels of polychlorinated biphenyls (PCBs) has been shown to cause impaired learning. Anti-thyroid effects have been noted in animals and humans and animal studies have revealed evidence of altered neurotransmitter and neuronal receptor levels, which may be primary or secondary to thyroid effects. The following is a list of POPs found in the food chain. (Don't try to remember these).

- Atrazine – an herbicide

- Dichlorodiphenyltrichloroethane (DDT) – a pesticide no longer used in North America. Interestingly, this compound is endorsed by the World Health Organization and could wipe out malaria
- Diethylstilbestrol (DES) – a synthetic estrogen
- Dioxin – industrial by-product of incineration, now in the food chain.
- Methoxychlor – a pesticide
- PCBs – no longer manufactured but still in transformers, electrical components and the food chain
- Vinclozolin – a fungicide

Sampling of foods reveals that organic foods contain pesticide residues 23% of the time vs. non-organic foods which contain them 73% of the time. According to a recent study, farmed salmon has high levels of PCBs.[53] Other pollutants such as mercury have been found in larger free range fish. The question is, do those levels pose a threat to humans? Very small amounts (nanograms) of the toxins are in many foods; we cannot stop eating all of them. While pregnant women, women of childbearing age and children should avoid high risk foods, we should not lose perspective; the benefits of fish (because of the omega-3s) outweigh the risks of poisoning. The following summarizes bad nutrition and high risk groups:

Unhealthy Foods

- Alcohol
- Candies – refined sugar, pigments
- Coloured foods – from unregulated sources
- Fast foods – saturated fats/trans-fatty acids
- Long shelf-life foods – industrial preservatives
- Sodas & soft-drinks – sugar, caffeine

Common deficiencies

- Ascorbic acid (vitamine C)
- Calciferol (vitamine D) – not found in food
- Calcium / magnesium
- Cyanocobalmin (B12)
- Folate (folic acid)
- Iodine
- Iron (protein and vitamin C are needed to absorb iron)
- Selenium

Common excesses

- Alcohol
- Fats
- Protein
- Salt
- Starch
- Sugar

Dangers – common poisons

- Alcohol – one drink is enough
- Recreational/addictive drugs
- Lead – paint, food colouring from unregulated sources
- Mercury – some fish, medicinals from unregulated sources
- Nitrates (40% of tinnitus)
- Persistent Organic Pollutants (POPs) – mostly from large fish and farmed fish

High risk groups

- Alcoholics
- Anorexics & bigarexics
- Children & teenagers
- Crash & other fad dieters
- Elderly
- Fast lifestyle; fast food consumers
- People on restricted diets – salt, fat, gluten, lactose, etc.
- Pregnant women

The elderly often have undernutrition. Improvements in health span and increase in life expectancy are clearly related to improved nutritional status.

Good nutrition

Earlier I quoted Hippocrates, the father of Western (allopathic) medicine, who said, "What is good for the heart is generally good for the brain". He was also the one quoted to say, "Let food be your medicine, and medicine your food". He learned that from the Chinese who traditionally made no distinction between foods, herbs and medicinals that might be prescribed as our doctors prescribe drugs.

Nutritionists and government guidelines tend to rely on what has been dubbed Recommended Daily Allowance (RDA) of caloric intake and nutrients. That concept remains highly controversial because it makes a number of assumptions and applies these in a blanket fashion to all of the population.[54] Furthermore, the derivation of the RDA values was never based on experimental (scientific) medicine – it never was and still is not evidence based! The RDA had its roots in efforts to prevent serious malnutrition in WW II soldiers receiving daily rations. The whole effort was guesswork, not informed by evidence-based medicine, and it was really meant to protect soldiers who were in combat for the short-term.

Problems with the RDA

As stated, the RDA guidelines are not based on long-term studies and rest on some erroneous assumptions. The RDA assumes that:

- All people have a balanced, omnivorous diet
- Each person's activity level is fairly high and uniform
- Nutrients are equally well absorbed by all people
- Requirements do not vary with age, gender or according to environmental factors

In addition, the recommended caloric intake for adults is overestimated: 70 kg. male, age 19-24 = 3,000 kcal /day; 51 kg. female, age 19-24 = 2,100 kcal /day. That formula will lead to relentless slow weight gain which will actually kill you. It is a good thing that people never took the RDA too seriously. Forget the RDA!

Since the end of the primitive days of the RDA era a whole new concept has emerged. Now we speak of the Optimal Daily Allowance (ODA).[55]

The following formulas have been derived to calculate daily caloric requirement taking into consideration gender, age, height, weight and activity level. This is the Harris-Benedict Energy Equation:

Males: 66 + (13.7 x wt.kg.) + (5 x ht.cm.) – (6.8 x age x activity factor)
Females: 65.5 + (9.6 x wt.kg.) + (1.8 x ht.cm.) – (4.7 x age x activity factor)

The above formulae assume that the subject is healthy. Activity factors change with various conditions as follows:

Physical Activity Levels:
Mild = 1.1
Moderate = 1.2 - 1.3
Heavy = 1.4 - 1.5

Stress Levels:
Mild (illness) = 1.1
Moderate (illness, surgery, infection) = 1.2 - 1.3
Severe (major surgery, burns) = 1.4 - 1.5

Clearly, the ODA is a better concept but, except for caloric intake, it has yet to be defined for various nutrients. We do have some very good guidelines, however. As we age, we need less food; normal healthy people eat less protein, carbohydrates and less fat. This observation fits well with the fact that after about age 54 (average for men and women) we slip into negative protein balance.[56] That is a menopausal change, entirely hormonally induced and cannot be altered in any way by increasing protein intake. As the protein matrix of the body (including bones), thins out, it is better to have less fat. Only moderately heavy regular exercise can slow this degenerative process and we will discuss the topic in the next chapter. It is natural and desirable to lose weight, very slowly and gradually as we age – as long as we maintain some fatty reserve to avoid putting caloric demands on non-fatty tissues.

What is optimal nutrition as far as the brain is concerned? A scientific analysis of many research projects on diet and severe NCI (dementia) reported at the 8th International Conference on Alzheimer's and Related Dementias, held in Stockholm (July, 2002), came to the same conclusion as the old Hippocratic doctrine. The findings were that a diet rich in fruit, vegetables, vegetable oil and fish offered maximal protection.[57] Why? The answer lies in the fact that the healthy brain diet is the same as the healthy heart diet as defined by the Canadian and American Heart Association.[58] It is essentially the Mediterranean diet that protects against atherosclerosis and blood clotting both in heart and brain blood vessels.

Features of the Mediterranean diet:

- High folic acid content from vegetables depress homocysteine levels and are believed to be an important factor in epithelial health inside blood vessels.
- Low animal fat levels help keep cholesterol down.
- High vegetable oil (olive oil) intake depresses 'bad cholesterol' which is an important factor in the development of atherosclerosis.
- High vitamin C and other antioxidants from fruit protect cells from oxidative damage.
- High levels of omega-3 fatty acids from fatty fish decrease platelet adherence and clotting tendency, thereby reducing risk of stroke.

- High N/C ratio; more nutrients and less calories than the regular balanced diet suggested by the old-fashioned 'food pyramid'.
- The Mediterranean diet is essentially the recommended 'food pyramid' of ten years ago turned upside down.
- The Mediterranean diet yields much vitamin C to aid in collagen/connective tissue formation

Healthy and Unhealthy Food Pyramids:

The name came from an imaginary triangle (pyramid) devised by nutritionists, indicating how foods should be distributed in our daily intake. We don't need to construct a 'pyramid', merely a list will suffice. The outdated list indicated that we should eat mostly starchy foods and carbohydrates, followed by meat and fatty foods, then smaller amounts of processed foods, even smaller amounts of sugar and sweets, followed by decreasing amounts of vegetables and grains, and finally a bit of fruit, nuts and seeds. The above is a 'meat and potato' diet with a bit of veggies and dessert! The newer, healthy list suggests almost the reverse. Today we recommend that most of the food should be vegetables, followed by lesser amounts of fruits (a recent study suggests not 5, but 7 servings of vegetables and fruits per day[59]), beans and legumes, followed by some grains and nuts with additional, smaller portions of fish and dairy products, a bit of poultry or eggs, a small amount of vegetable oils and finally, if you must, tiny loads of processed foods, sweets, beef, cheese and milk! What we recommend today is essentially the Mediterranean Diet. Don't get me wrong, you can get fat and unhealthy on the Mediterranean Diet if your caloric intake is higher than your expenditure. In any case, you will have less atherosclerosis on the healthy option.

The American Heart Association diet of fatty fish, vegetable oils, vegetables, cereals and fruit is about the same as the Mediterranean Diet. The main benefits are a reduction in trans-fats, animal fats (saturated fats) and an increased intake of vitamins, minerals, fiber and antioxidants.

We have all heard and read about antioxidants and antioxidant capacity. A large component of that protective effect comes from physical fitness to be covered in the next chapter. However, dietary sources remain very important. Antioxidants combat free radicals. Free radicals are 'rogue' oxygen atoms that destroy parts of cells causing cell death. As we age, dietary antioxidants become more and more important. We all heard about vitamin E and beta-carotene acting as powerful antioxidants and then research has shown that taking large amounts of these supposedly protective substances had no benefit at all and they even increased the risk of disease. What happened? The story is all the more intriguing because hardly anyone understands the issues involved. For example, the American Association of Neurologists have recommended 2,000 IU of vitamin E for over a decade for the treatment of AD and despite new research proving it worse than useless, it is still in the official treatment guidelines. I will now try to explain in plain language why these marvelous antioxidants backfire when taken in purified form and in large doses. There are many, perhaps tens of thousands, of different types of anti-oxidant molecules. These belong to families of naturally occurring compounds which include carotenoids, tocopherols, polyphenols, cannabinoid-like compounds and flavonoids, just to name a few. Those substances are responsible for the colouration in fruits and vegetables. A tomato may contain a thousand different lycopenes that belong to the family of carotenoids. Those antioxidant molecules are very specific and protect particular other molecules from rogue oxygen atoms. There is no such thing as an ultimate antioxidant acting in a blanket fashion. Yet, when the health and food industry (and the consumer) latches onto a 'new' antioxidant, they react in an all or nothing fashion. Usually, what is sold is what is easily manufactured in large amounts. The resultant 'purified' simple forms prevail. Hence we had alpha-tocopherol, instead of mixed tocopherols and beta-carotene instead of mixed carotenoids. It was discovered that when one consumes a single molecular version of a family of antioxidants, that single compound can competitively inhibit the absorption of all the other compounds of the same family! In other words, taking alpha-tocopherol blocks all the other tocopherols (beta, gamma, etc.) and could produce a relative vitamin E imbalance (insufficiency or deficiency). Likewise, taking large doses of beta-carotene blocks the other carotenes producing a relative carotenoid deficiency. Imagine what would happen if we did this with all antioxidants! Fortunately mother nature is too complex for us. Our brain, being the most complex entity in the universe that we know of, can overcome some of these insults without drastic reactions. The best dietary antioxidants remain complex foods, not drugs, which are rich in flavonoids, carotenes, tocopherols, lycopenes, polyphenols and many more compounds. That is why we will never be able to replace a raisin or even a grape seed with a synthetic chemical offered as a pill.

Total Radical-trapping Antioxidant Potential (TRAP) values

Consider the following study, the Oxygen Radical Absorbing Capacity (ORAC) Study.[60] It was a randomized, placebo controlled, cross-over study of four female and four male athletes during triathlon, twice, two weeks apart. They were given 170 g of raisins or equivalent calories in glucose, before and during competition. Results: Raisin group had virtually no oxidative damage in their cells. The glucose group did.

Foods can be classified according to their Total Radical-trapping Antioxidant Potential (TRAP) values. Single antioxidants do not have high TRAP values. Antioxidants act synergistically, reactivating each other to yield maximal effects.

A diet rich in naturally occurring antioxidants is much better than hundreds of antioxidant pills. The only trade-off is that fruits, especially juices are high in sugar content. Don't forget that many juices have sugars like fructose or sucrose added to them, which, like pure glucose, only adds to the oxidative stress.

Foods with high TRAP values:

- Fruits
- Berries
- Seeds
- Vegetables

Fiber

What is the relationship between fiber intake and brain health? The role of fiber in health, particularly colonic health has been well publicized, but what about the other end of the body, the brain? Fiber does a lot more than provide roughage. It degrades in the gut to produce natural substances with anti-inflammatory and anti-hypertensive properties. The result is a protective effect against cerebrovascular disease and stroke.

Fats & oils

No discussion on brain health would be complete without at least mentioning fats and oils. Brain tissue is at least 70% fat! Yes, the brain is made mostly of fat because the nerve cell membranes and their coverings (myelin – the white matter of the brain) are all made of molecules of fatty substances. It follows that children with rapidly growing and developing brains need fat – the right kind of fat. That does not mean they have to become unfit and overweight or obese. It is not surprising that human milk is very high in fat content – higher than cow's milk. (Whale milk is so high in fat content that it is almost solid, like soft butter.) Mother's milk makes you smart. For a higher IQ, more appears to be better, ideally for 12 months or more.

Vegetable oils are extremely important in cooking. They are extracted either from seeds (such as soya and sunflower seeds) or from fruits (such as olives and nuts). Sesame and olive oils have the oldest origins. Records show that both were used by the ancient Egyptians. The ancient Greeks used olive oil. Most vegetable oils are low in cholesterol, being made up of monounsaturated or polyunsaturated fatty acids. Others, such as coconut and palm oil, contain almost as much saturated fatty acid as animal fats. The less saturated an oil/ fat is, the more unstable and the more likely that it will turn rancid even at room temperature. Heating an unsaturated oil like olive oil, breaks it down into deleterious, even cancer producing components. Unsaturated fats (say olive oil, flax seed oil, fish oils) have a 'bad cholesterol' lowering effect but cannot be used for cooking, particularly frying, because of the high temperatures involved. Unsaturated oils should be kept in a cool dark place (fridge). Saturated fats are more stable and can be stored more easily and can withstand high heat (animal fats, bacon fat, peanut oil, etc.) but they tend to contribute to atherosclerosis by increasing the 'bad cholesterol' in the blood.

Consumption of fish and n-3 fatty acids and risk of incident Alzheimer's disease (AD)

A prospective study was conducted from 1993 through 2000, of a stratified random sample from a geographically defined community. Participants were followed up for an average of 3.9 years for the development of AD. A total of 815 residents, aged 65 to 94 years, who were initially unaffected by AD completed a dietary questionnaire on average 2.3 years before clinical evaluation of existing AD. Result: Consumption of the n-3-polyunsaturated fatty acids and fish was associated with reduced risk of incident AD in this large prospective study.[61] Persons who consumed at least one fish meal per week had 60% less risk of AD than did persons who rarely or never ate fish. Of the marine n-3-fatty acids, only docosahexaenoic acid (DHA) was protective against the development of AD. Intake of alpha-linolenic acid was also protective, but only among persons with the APOE-e4 genotype. The study has a number of strengths that support the validity of the findings. Protective associations were observed for DHA, a major component of brain phospholipids, with fish as its primary food source, and with alpha-linoleic acid, which is largely obtained from vegetable oils. The strength of these associations makes it unlikely that the findings are due to chance. Those associations held even after adjustment for education and other important risk factors, including cardiovascular conditions that could potentially account for the observed relative risks. Conclusion: Dietary intake of n-3-fatty acids and weekly consumption of fish was associated with a reduction of risk of incident AD.

Phospholipids and the brain

DHA is a primary component of membrane phospholipids in the brain. High levels of DHA are found in the more metabolically active areas of the brain, including the cerebral cortex and its cells. Fish is a direct dietary source of preformed DHA. In addition, DHA is synthesized endogenously through a process of desaturation and elongation of its precursor n-3-fatty acids, alpha-linolenic acid and eicosapentaenoic acid molecules. The above fats are more slippery and less likely to clump. In laboratory studies, animals fed diets enriched with n-3-fatty acids had better regulation of neuronal membrane excitability, increased levels of neurotransmitters, higher density of neurotransmitter membrane receptors, increased hippocampal nerve growth, greater fluidity of synaptic membranes, higher levels of antioxidant enzymes, decreased levels of lipid peroxides, reduced ischemic damage to neurons and increased cerebral blood.[62] In short, they had healthier brain tissue. In behavioral models, animals fed diets enriched with n-3-fatty acids had superior learning acquisition and memory performance over animals fed control diets. A human case-controlled study reported that n-3-fatty acid levels in plasma phospholipids of patients with AD were 60%-70% of levels found in age-matched control subjects.[63] Other studies found that fish consumption was inversely associated with risk of incident AD.[64]

Are drugs the only treatment for high cholesterol? No. Many people can lower cholesterol with diet changes alone. A study in the *Journal of the American Medical Association* (July 23, 2003) showed that a combination of cholesterol-lowering foods is as effective as statin drugs in lowering both blood cholesterol and inflammation.[65] The subjects who were fed a diet that included plant sterol margarine, soybeans, almonds, psyllium fibers, barley, oats, eggplant, okra and many other vegetables had the bad LDL cholesterol levels drop by 29 percent compared to 31 percent in the statin drug group. A cholesterol-lowering diet appears to be completely safe, while statin drugs may have side effects (muscle pain, depression, diabetes). However, statins are efficacious in lowering 'bad cholesterol' for those who cannot have the discipline or opportunity to pursue a strict cholesterol lowering diet – most people. Athletes and serious exercisers train by taking a hard workout that makes their muscles sore on the next day, take easy workouts until their muscles feel fresh again and then take another hard workout. Statin drugs make it difficult to maintain a good training program because they can delay muscle recovery. However, the overall benefits of statin therapy appear to outweight the side effects, which in most cases are on par with placebo responses.[66]

The bottom line is that consumption of fruit and vegetables is positively related to low levels of LDL (bad cholesterol) in both men and women.

THE LEAST YOU NEED TO KNOW

- Bad nutrition is rampant; health = high nutrient calorie ratio. (H = N/C).
- Mediterranean diet, variety and exercise are best.
- Restricted diets must have supplementation.
- Deficiencies can be avoided or treated.
- Herbal preparations may have efficacy but also dangers.
- Foods, herbs and drugs can interact dangerously.
- Foods and herbs can have toxic impurities.
- The right kind of fats are good for the brain.

START NOW!

- Read and learn about proper nutrition.
- Avoid empty calories – if overweight, eat less – micromanage meals.
- Analyze diet – see a dietitian. Don't crash diet - take a crash course in nutrition instead.
- Learn about toxicity of certain foods, herbs, drugs. Read labels.
- Emphasize quality and variety instead of quantity of food.
- Don't eat fast. Don't eat to be full. Look for twinges of fullness instead of hunger.
- If you use alcohol, use it in strict moderation.
- Don't shy away from obesity talks – counseling works.
- Ascertain your BMI. Once you have your BMI, use the table in this chapter to ascertain your risk of illness. Reduce your BMI as needed!

Your Healthy Brain: www.HealthyBrain.org

CHAPTER 9

PHYSICAL EXERCISE: THE BRAIN'S FOUNTAIN OF YOUTH

Above all, do not lose your desire to walk. Every day I walk myself into a state of wellbeing, and walk away from every illness. I have walked myself into my best thoughts, and I know of no thought so burdensome that one cannot walk away from it.

— Soren Kierkegaard

When we need these healing times, there is nothing better than a good long walk. It is amazing how the rhythmic movements of the feet and legs are so intimately attached to the cobweb cleaners in the brain.

— Ann Wilson Schaef

In this chapter I'm not talking about mental calisthenics, endurance reading or strength training for your vocabulary. That type of exercise will come in the next chapter. I'm talking about walking, running, weight lifting, sit-ups and the rest of the things that you do to raise your heart rate, work your muscles and improve your aerobic capacity.

The title of this chapter is not an exaggeration. Exercise keeps the brain young and healthy. The reasons for this are both extremely simple and extremely complicated. The simple reason is that exercise improves the flow of blood to the brain and the amount of oxygen available to the brain. That improves the brain's metabolism and allows it to work more efficiently and effectively. The complicated reason is that exercise sets in motion a series of chemical and hormonal activities that stimulate brain growth.

Prescription for exercise

I once had a patient who will always stand out in my mind because she was exceptional in one respect – treatment adherence. Most patients do not follow a doctor's orders fastidiously but Mrs. Wong was different. She was referred to me because of severe depression that beset her after a great deal of family hardship and stress. She was 80 years old and fit as a fiddle. Prior to her illness she enjoyed walking for miles every day. By the time I saw her she was out of condition and suffered from extreme fatigue. She was on a high dose of antidepressant but her depression remained refractory and she spent most of her time in bed. She had given up all previous activities like walking, playing piano, reading and interacting with friends and family. She was very intelligent, had good insight and desperately wanted to get better. She recognized that if her trend continued she would further deteriorate, both physically and mentally. She would become moribund and die. I reduced her antidepressants because of side effects. I insisted that she exercise. She agreed that this was a good idea but she was so weak and tired all the time that she could not exert herself in any way for more than a few minutes – she was vegetating. "Ok", I said, "then you will walk for three minutes per day". She thought this was silly compared to what she wanted to do but we discussed the plan, made some simple charts and started an exercise program. We increased the walking time by about five minutes per week. This took a huge effort at first but she stuck to the plan. Six weeks later she was walking half an hour per day – still much less than what she used to do. She was feeling better and her concentration improved. She could read again and she even played the piano occasionally. Family noticed that she was in a better mood. She stuck to the program and four months later she was back to her old self – in 'complete remission', as doctors would say. By that time she was fully active. She decided to walk everywhere instead of driving (although she could drive very well) and she could walk for hours without difficulty. I often wonder why doctors don't treat exercise as a drug and prescribe it the same way, particularly for the elderly. When physicians prescribe drugs we follow the rule 'start low, go slow, but don't stop'. I believe in prescriptions for exercise that follow those same principles.

If you are like most people you either exercise regularly (some of us) – or at least think about exercising regularly (the rest of us). One of the reasons that exercise is on our minds is that the medical community and society in general have raised our awareness of the health benefits of exercise. Another one may be that the body actually craves exercise. While we don't know if exercise prevents diseases from occurring, we know with certainty that if you exercise regularly you will feel better, have better cardiovascular status and have a healthier immune system. There is a pretty strong consensus that exercise is good for us – it certainly increases health span and probably life span.

Have you ever thought about how exercise affects your brain? If you haven't, it's time to start. Exercise is one of the most powerful tools available to us as we work to optimize our brain capacity and increase our brain's staying power. The following table is a short list of what brain chemicals do after being stimulated by exercise.

Exercise increases all of the following:

- Branching of brain cells and making new connections (neuroplasticity)
- Birth of new brain cells (neurogenesis)
- Formation of very small blood vessels (angiogenesis)
- Hormonal support for glial cells (the cells that feed the neurons)

Before getting into the science of exercise and the brain, let's take a quick review of exercise and healthy living, starting with the consequences of inactivity. The disorders in the table have been carefully chosen from the much larger list of medical conditions that can occur as you grow older. Not all medical problems can be prevented by exercise but the ones listed below are strongly linked to inactivity.

Diseases associated with physical inactivity:

Hyperlipidemia*	Hypertension*	Stroke*
Chronic stress*	Obesity*	Diabetes*
Depression*	Chronic hormonal disorders*	Sleep disorders*
Heart disease*	Osteoporosis	Colon cancer

* Note that the diseases marked with an asterisk are major risk factors for NCI.

The above is a very scary list and I am going to make it even scarier by adding another item to the list: brain disease. As we study people and aging more closely, it becomes clear that those who remain physically active have better health status than those who become or remain inactive as they grow older. One of the findings is that the rate of NCI is lower in groups that remain active than those who are inactive.

Longitudinal research projects follow the same group of people for long periods of time. These studies allow researchers to closely examine people over time and give us a better idea of the long-term effects on health of such things as exercise, diet, etc. Much of what we know of health over the life span comes from longitudinal studies. The message from modern clinical research is that you neglect your body at your own cost. The problems listed above are difficult and costly to treat as they lead to disability as well as untimely death. Unfortunately, the cost of managing the problems of the inactive spreads to the rest of us. Canadian figures suggest that the cost of couch potatoes to the health care system exceeds billions of dollars annually.[1] In the USA, the figure would be closer to tens of billions annually. So if you won't become active for your own sake, please do it for the rest of us. We can't afford to pay the bills anymore!

"We know that if everybody exercised a few hours a week, Type II diabetes would be virtually non-existent" – Dr. K. Goodrick, Baylor College of Medicine, Houston, 2001.[2]

The Dallas Bed Rest and Training Study 1966-1996 was conducted to examine the effects of age on exercise tolerance and fitness.[3] The researchers measured the effects of two six-month endurance training programs after a 20-day bed rest period, 30 years apart. Conclusions:

- There is an age-related decline in aerobic power/endurance. (Decline in the body's ability to extract oxygen i.e. to receive, take up and use oxygen.)
- A six month aerobic exercise program can reverse a 30-year age-related decline.
- Older persons who have failed to maintain fitness over time can benefit from an exercise program.
- Three weeks of bed rest deconditioning has a more profound negative impact on physical work capacity than 30 years of aging.

Unfortunately, we know from psychological literature that presenting you with scary information will have very little long-term impact on you. If this tactic worked, there would be less crime, fewer accidents and a lot fewer bankruptcies. The reality is that most of us easily trade the immediate pleasure of our sedentary leisure time for the remote possibility that all will be better later. This is a costly fantasy to uphold. If we are sedentary, we will be more likely to fall ill – sooner rather than later. That's reality – that's life. We know that you are more likely to change your behaviour if there is something in it for you – right now.

This chapter lays out some of the immediate benefits that will come to you if you start to become active. First I want to expand on the idea of activity. I'm not a personal fan of strenuous physical exertion, although it is recognized that some people are. Fortunately, the studies of fitness and health do not prove that you need to be an athlete to be healthy. In fact, over-exercising can be as bad for the body as under-excising.[4]

From my point of view, being active doesn't mean working out at the gym everyday, buying expensive exercise equipment or getting into bodybuilding.

Being active can be as easy as walking 20 minutes every day, joining a yoga or Tai Chi class or having a swim in the pool. What is important is that you avoid the inactivity trap and do something on a daily basis, even if that is walking up the stairs instead of taking the elevator. Oh, and one more thing! The exercise you do must be voluntary and if not downright pleasurable, at least refreshing. Forced, compulsory, excessive or compulsive exercise may be so stressful that it may not be worth doing. You will learn more about the harmful effects of stress in Chapter 12.

It is extremely important to understand that commitment to exercise is a lifelong affair. To achieve good health benefits, exercise must be regular and sustained over long periods of time. This does not mean, in fact it must not mean, that you have to make a huge effort every time you exercise. It should be a habit. If it is a habit, it will feel uncomfortable and stressful to not exercise. It will be easy. Don't get stressed out over it. You can experience benefits very quickly, but lose them just as quickly. What seems to work best from a health perspective is a change in lifestyle that takes place gradually, becomes a habit and continues over the years. This is why I advocate easy to do exercise like walking and low impact activities like Tai Chi and yoga. They are not only possible for people of all ages to do, they can be much more easily built into long-term routines than trips to the gym. Even short periods of exercise like taking a brisk walk for 15 minutes two or three times a day is very beneficial.

Older adults show a real decline in brain density in white and gray areas, but fitness actually slows that decline. Don't be so surprised – the same applies to muscles and bones. The body, which includes the brain, needs physical exertion and execrcise. Researchers unanimously agree but more research is needed to precisely understand those effects. The bottom line is that fitness, as it interacts with age, has strong positive effects. A recent American study used modern brain scan techniques to study how exercise physically effects the brain.[5] In the study, researchers used MRI to analyze three-dimensional brain scans of 55 well-educated adults aged 55 to 79 who ranged in fitness from sedentary to competition-ready athletes. The research scientist, Dr. Arthur Kramer of the Beekman Institute at the University of Illinois found that older adults who participated in exercise had a higher brain density (more brain cells) than did adults who did not exercise. Dr. Kramer concluded that fitness interacts with age in a very positive way to slow or reverse age-associated decline in brain structure. The benefits of aerobic exercise extend well beyond cardiovascular health markers and can effect brain health as well. Conclusion: you need to start early and keep on exercising if you want your brain to remain fit and healthy throughout adult life.

Having said all of that, I want to direct your attention to the benefits of physical activity on the most important organ – the brain. There are both direct and indirect, both physical and psychological effects on the brain when you participate in regular physical activity. The direct physical effects are particularly dramatic and exciting so we will start with them.

Direct effects of physical activity on the brain

The first direct physical effect, and one that you can easily perceive, is a release of endorphins. Endorphins are the 'feel good' hormones, actually natural internal opiates, released when the body is exercising. Many of you will have heard of 'runner's high' the phenomenon that occurs when long-distance runners experience a state of euphoria. What many people don't realize is that you don't have to sustain extreme effort. Moderate exercise is sufficient to get the endorphin system going and this contributes to the good feeling and euphoria that is associated with regular exercise. The endorphins also reduce pain sensitivity. Other brain chemicals called monoamines are released rapidly and sustain their effect for about twelve hours. Adrenaline, noradrenaline, serotonin and dopamine are all whipped up and flood the brain giving it a natural boost of energy and euphoria. Don't forget that those brain chemicals are the ones increased by modern antidepressants. Actually, heavy exercise is as potent in upping brain concentrations of those natural nerotransmitters than the pills – exercise can be prescribed as an antidepressant.[6] Unfortunately people with depression usually cannot sustain heavy exercise, otherwise many of them could cure themselves without drugs. Similarly, there are many people with physical disabilities who cannot achieve even moderately heavy exercise. For those who are able, physical exercise can be an effective therapy against cognitive decline, depression, Parkinson's disease and a few other ailments, some of which are in themselves risk factors for degenerative brain disease.

Indirect effects of physical activity on the brain

The so-called indirect effects are slower, can't be felt immediately and are more sustained, but no less important. When you exercise regularly you initiate two related processes. 1) You produce physical fitness, beefing up the heart, lungs, arteries and small arterioles that deliver blood to the brain with much greater efficiency. 2) You stimulate the important inner brain processes: neurogenesis and angiogenesis. Neurogenesis is the process of creating new brain cells. As you learned in an earlier chapter, the brain has the capacity to grow and neurogenesis is the process of growth. If you exercise on a regular basis you improve your brain's health by stimulating the process of neural growth. Angiogenesis is the process involved in healing wounds by restoring blood flow to tissues and growing new blood vessels. Angiogenesis is particularly important in the brain as the maintenance of the vascular system is crucially important to keeping your brain well supplied with fresh, oxygenated blood. (Remember it uses one-third of the blood supply and 25% of the energy of the whole body.) There is plenty of research to show that exercise changes the brain. Dr. Brian Christie, a neuroscientist at the Brain Research Centre at the University of British Columbia Hospital and the Department of Psychology at UBC found that exercise improves performance in cognitive tasks and leads to functional changes in brain regions.[7] In a second study, they showed that new brain cells are formed in the hippocampus after exercise and that existing neurons have longer dendrites, the branch tendrils that allow communication between cells and can communicate with one another more efficiently. "We originally showed, in 1999, that exercise could improve neurogenesis and synaptic plasticity in normal animals," Christie notes. "This new research is the first extensive quantification of how the brain changes following exercise".[8]

Exercise promotes brain health and function by keeping these two vital processes operating so that growth and healing occur. Over a lifetime, it is impossible to overstress how important this is in keeping your brain intact and robust. The stimulation of these two processes is also of critical importance in warding off diseases that may strike the brain. (More about this in Chapter 14 where we will look at some common disorders leading to brain disease). At a very basic physiological level, regular exercise produces fundamental and positive effects on your brain.

There is one more little secret. That pertains to what is known as *oxidative stress*, which acts on every cell in the body including in the brain. Oxidative stress degrades the brain, especially in an area called the hippocampus. (The hippocampus is dedicated to the learning and retrieval of new information). People who exercise regularly have a lot of oxygen molecules to deal with. Their bodies have to have a mechanism to keep oxidative damage to a minimum and they succeed by ramping up an inner, natural antioxidant system. In the prevous chapter on *Nutrition* we did discuss dietary antioxidants. Yes, they are important but the fundamental antioxidant capacity of the whole body is just as, if not more dependent on exercise related antioxidant capacity. Yes, as unbelievable as it sounds, it is a fact that people who exercise regularly have strong protective mechanisms in place to prevent damage even after insults to the brain. In other words, they have added protection against the oxidative stress not only after injuries (concussion, stroke, Alzheimer's, etc.) but also against the aging process itself. That is why physical fitness is the closest we can come to the *Fountain of Youth*.

Exercise offers protection against many diseases but some of the studies on Parkinson's disease are exemplary. In a recent pilot study (not a study of pilots but a small, preliminary study designed to determine if it is worth engaging in a large study), Annie D. Cohen, a doctoral student in the Department of Neurology and Center for Neuroscience at the University of Pittsburgh School of Medicine, examined the brains of rats that had been forced to exercise for seven days before receiving a toxin that normally induces Parkinson's disease caused by the death of dopamine containing neurons. They found that, compared to animals that had not been exercised, significantly fewer dopamine-containing neurons died. "Whereas a number of explanations could be offered as to why the exercised animals do so well, we have evidence that indicates it's because exercise stimulates production of key proteins that are important for survival of neurons", said the study's senior author, Michael J. Zigmond, co-director of the university's Parkinson's Disease Clinic.[9] While this work has yet to be applied to humans, it points to the importance of physical activity in promoting brain health. Parkinson's disease belongs to a group of conditions called motor system disorders. The four primary symptoms are: 1) tremor or trembling in hands, arms, legs, jaw and face. 2) rigidity or stiffness of the limbs and trunk. 3) bradykinesia or slowness of movement, and 4) postural instability or impaired balance and coordination. As these symptoms become more pronounced, patients have difficulty walking, talking, or completing other simple tasks. We cannot demonstrate the above protective mechanisms in humans because we can't see the brain cells before and after exercise. In animal models it is easy to sacrifice the animal and have a direct look at brain regions. Clinical observation and results of human longitudinal population studies are consistent with animal research[10] and I have little doubt that exercise has similar effects by similar mechanisms in all people.

Psychological effects of physical activity on the brain

In addition to the dramatic physical effects described above, exercise has psychological effects which are readily observed and which have been well documented in many studies. No wonder – if the brain improves physiologically, we expect a functional improvement as well.

The first and most immediate effect you experience when you exercise is a sense of well-being. People who engage in regular exercise consistently report that it makes them feel better in general.[11] This is the wonderful consequence of getting up off of the couch and starting to move around. If you want to feel better, get moving. Your brain will lead the way!

The second immediate benefit will be an increase in your overall energy. If you take my advice and follow a moderate regime, exercise will not tire you out. On the contrary, it will energize you both physically and mentally. The increase in mental energy will include an increase in mental efficiency. Studies show that when people engage in sustained exercise programs, their improved cardiovascular function is accompanied by improved performance on tests of mental efficiency. Exercise won't necessarily raise your IQ, but it will help you to perform at your highest level of cognition efficiency – without added effort.[12]

The third benefit is that you will feel better about yourself.[13] One of the well-documented effects of exercising is an increase in self-esteem. When people become active, they begin to think differently about themselves – in a good, healthy way. People report that they experience a sense of mastery and an improved self-concept; they feel more in control of their lives and better about who they are. If you stick to your activity program, the short-term sense of well-being that you experience will turn into an enduring improvement in your self-esteem. Psychologists call this 'healthy narcissism', but I call it *feeling good.*

Now you know that if you exercise your brain will be healthier, more resistant to disease, and more efficient. You know that if you start a program of mild to moderate regular activity you will feel better about yourself and will perform at a higher level on mental tasks. Make it your goal, your purpose to take the first step, however small, to start a regular exercise habit – like my patient Mrs. Wong.

Psychological resistance to exercise

Before you get started on your new program of healthy living, you need to look squarely at the factors that are likely to get in your way. The first problem: your accustomed habit. We all tend to be creatures of habit and many of those habits aren't in our best interests. For example we may have a habit of coming home from work and falling into a chair to watch the news. The first step to changing our habits is to admit that they are there. Make a list of some of your inactivity habits and then, beside each one, create the new activities that you want to turn into good habits. This takes time but in order for change to take place you need to do it. The brain, as complex as it is, perhaps because it is as complex as it is, requires routines and habits. Once you have established a new habit, it will carry you forward. A habit does not come about just because you make a resolution – it needs actual practice. If you have a sedentary habit you have probably practised it for a long time. It takes a little effort and about 30 days of practice to make (or break) a habit.[14] All the habits you have are there because you practiced them for at least a month or so – most of them a lot longer. A new habit can be formed by deliberately practicing the act for over 30 days. After that, the new behaviour will become automatic; your brain will have it built in and it will feel uncomfortable to stop. Just think how uncomfortable it would feel to stop brushing your teeth or to stop going to sleep at your usual time. That's because these are autonomous habits. If you make a habit of walking every day, you will feel uncomfortable and strange if you stop. When forming good habits, your brainpower is your friend.

The second problem is negative thinking. Like behavioural habits, negative thinking can become a way of life. We all accept the immense power of positive thinking – it can change your life, make you great, make you successful.[15] Well, the reverse is also true. Negative thinking can drive you into an abyss, make you passive and promote depression. Take a look at your own thought patterns and try to find the negative thoughts that stand in the way of becoming more active. Do you find yourself thinking that there isn't enough time, I don't like people watching me, I hate feeling uncomfortable and so forth? Every time that you find a negative thought substitute a positive one. If you find yourself thinking that you don't have enough time, substitute the thought that there is plenty of time. Really, you have plenty of time to take a walk. Treat your thoughts the same way as you are going to be treating your inactivity habits. Identify them, mark them down and then change them.

The third problem you will need to be aware of is stress. If you experience a lot of stress in your life, it is likely that you will be less inclined to engage in activity as the stress will wear you down and leave you feeling unmotivated. This is a dirty little trick that your brain will play on you. In fact, if you build mild to moderate physical activity into your life, you will not only experience less stress, but if you find yourself under stress you will be better able to endure it. When you engage in moderate physical activity, it actually can turn off the physical stress response in your body by providing a biochemical counterbalance. It can stabilize your heart rate, lower your blood pressure, improve your breathing, stabilize your stress hormone levels and generally calm you down. If you wish to combat stress and the consequent limitations it can place on your life, you need to recognize that physical activity is one of your most important stressbusters. (More on stress in Chapter 12.)

The fourth problem, and one that I cannot emphasize strongly enough, are cultural factors. By cultural I mean not only national differences, although we know from the Swedes how a nation's identity can be built around awareness of the need to be fit. Often smaller cultural factors, such as the social environment of your workplace or of your extended family can significantly impact your activity level. If, for example, your colleagues at work all sit in the lunch room and talk at noon, chances are that you will join them and lose the opportunity for an activity break. Similarly, if your entire family is into television and sedentary games, chances are that you will not buck the trend and head off for a bicycle ride. You cannot always control the culture, but you can make sure that it does not control you and stop you from achieving the level of activity you need to keep your body and brain healthy. Talk to those around you about the benefits of your approach. Chances are that at least some of them will be interested in keeping their brain healthy and join in with you. Make the culture work for you. It may take courage to buck a trend but if your choice is healthy, others may jump on your bandwagon to support you.

The fifth problem is misconception about fitness. In North America we have illusions about what is fitness and what is beauty. An unhealthy perceived relationship between fitness and beauty is reinforced by a huge commercial industry with gyms, trainers and space-age exercise devices along with a barrage of claims that exercise involves 'buns of steel' and 'six-pack abs'. In fact, medical fitness has nothing to do with the size and shape of your muscles. The Healthy

Brain approach has much more to do with the quality and harmonious function of the organ systems of the body. Body mass, in fact, is not that closely related to fitness. You need to be aware that to be fit, you only need to engage in regular, light to moderate activity. Walking is one of the most natural and beneficial activities known to man and the only reason it is not advertised on all the major channels is that it is free. Don't buy into the 'fitness-beauty' myth. Exercise for your health, not the mirror. There is nothing more beautiful than a healthy person. What is more wonderful than a happy old face?

This discussion leads to the ultimate question: What activity program is right for me?

My way to exercise

There is no single answer to this question. The main thing is to get going. Everyone is different; therefore, exercise programs must be unique for each person. People in wheelchairs exercise. People with arthritis exercise. People with heart disease exercise. Design a gradual program for yourself. Treat it like a powerful drug from the Fountain of Youth – start low, go slow, don't stop! If you need the help of a fitness trainer or physiotherapist – go for it!

There is a s self-guided program that was developed, many years ago, for pilots by the Royal Canadian Air Force. Don't let this intimidate you. It is very easy to start, requires no special expertise and is inexpensive; available at libraries, bookstores and on the internet free of cost. You can use this to get a routine going. It will gradually put you into excellent physical condition. Get 5BX for men or 10BX for women. Still available – highly recommended.

Manpo-Kei is the art and science of step counting. It is a way to be naturally active. In Japan, where there are many clubs promoting programs using pedometers to log walking and hiking, Manpo-Kei has become a slogan. The pedometer has become a common household item. The goal is to walk up to 10,000 steps a day.

You can devise an exercise program that does not involve walking. Look them up on YouTube or Internet. Floor exercises, water exercises (hydrotherapy) and even exercises that can be done in bed are freely available (mind you, that breaks the rule about *sleep hygiene* to be covered in Chapter 11 on Sleep.)

There are ways to increase daily activity that require no special time commitment. In some situations hey may actually save time. You must explore such ideas. Whenever possible, walk – don't drive. Ride a bike. Take the stairs instead of the elevator. Analyze your daily activities and travels and increase physical activity bit by bit. Try working standing up instead of sitting. Nothing is worse for your body than sitting!

Fitness is more important than BMI. A person who is lean and sedentary is at a higher risk for stroke and heart attack than an overweight fit person.

THE LEAST YOU NEED TO KNOW

Regular, mild to moderate exercise will improve:

- Brain function – concentration and memory
- Quality of life
- Feeling good and self esteem
- Health span
- Life span
- Brain health

Exercise also:

- Protects against the ravages of aging and disease, particularly brain disease.
- Helps keep your brain cells young.

There is no legitimate excuse for avoiding the creation of the exercise habit.

START NOW!

- Get going – start walking 15 minutes every day.
- Start an exercise program like RCAF 5BX, 10BX plan or Manpo-Kei (10,000 steps) program – these are easy to start and require no need for a professional helper.
- Increase exercise gradually – start low, go slow, continue but avoid distress.
- Use an exercise log to stay focused.
- Start a journal or logbook to reduce sedentary activities.
- Remember: regularity leads to habit formation after about 30 days.
- Keep a positive attitude; focus on health span and good self esteem.
- Don't be shy about using a fitness club, personal trainer or physiotherapist if that is what you need.

www.HealthyBrain.org

CHAPTER 10

MENTAL ACTIVITY

No wise man ever wished to be younger.
— Jonathan Swift

I am always amazed that how I start the day affects what transpires during the rest of it. This is logical. Thoughts reverberating around in the cerebral cortex generate feelings and feelings are more enduring, although they too will dissipate. Hence, brain activity will generate a set of thoughts and feelings that affect output to deeper emotional centers, which will in turn generate behavioral patterns and expectations and external results, predetermined by previous activities. I make this statement to emphasize that what goes on in the brain matters a great deal. My thoughts create attitudes, attitudes create particular expectations and moods; those translate into corresponding actions or non-actions, which translate into a set of circumstances. That is what I have to live with and I am pretty sure the same thing applies to everyone.

In this chapter I explore mental activity or brain activity and how exercising the brain through various types of stimulation, purposeful or otherwise, affects its development, general functioning and even degeneration.

I remain fascinated by the well-known research into the exciting area of brain stimulation and enhanced brain longevity. It is of special interest that animal studies and observations on human development and aging are consistent – they support the same conclusion: *use it or lose it!* As a corollary, I posit that the way we use the brain has important consequences. Mental activity and optimal stimulation are very important not only for brain development but also for the maintenance of baseline levels. Lack of mental activity is bad for the brain. It appears that a lot of people do not know what 'mental activity' and 'mental stimulation' really mean.

We will now review some simple definitions which will make the next paragraphs more comprehensible. For our purposes, *mental stimulation* will refer to input to the brain that the brain has to process mainly through the cortex of various regions. Of course, the brain has to process everything – huge amounts of data from all parts of the body and blood. However, we will not consider that 'mental'. We will focus on the basic senses (sight, hearing, smell, touch, taste, balance) and intellectual activities (thinking, speech, perception, etc) that we readily recognize as a mental or psychological information. So much for input; what about output? That we will call *mental performance*. Mental stimulation and mental performance translate into *mental activity* – the topic of this chapter. How does mental activity affect the brain?

When it comes to brain stimulation there is an optimum level. (We will see a similar relationship when it comes to stress and brain function in Chapter 12.) Let's jump right in and look at some interesting research – research that is highly repeatable. The relationship between *Mental Stimulation* and *Mental Performance* follows an inverted U pattern. Too little mental stimulation results in poor performance. The right amount yields optimal (best) performance (at the top of the inverted U) and too much stimulation also gives poor performance.

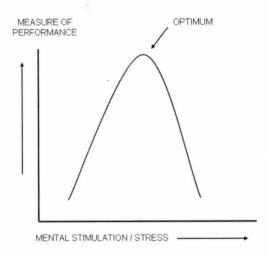

RELATIONSHIP BETWEEN
MENTAL STIMULATION and PERFORMANCE

For example, if people are asked to recognize speech, they can only do the task if the input is near optimal. If the input is too quiet, slow or fragmented, recognition will be poor. If it is too loud and too fast, performance will again suffer. Thus, *mental activity* may be experienced as frustrating (can't hear), rewarding (comprehensible, enjoyable) or distressing (obnoxious noise). At higher levels of complexity, the mental activity will also be registered as irrelevant or meaningful and that too will affect performance. When the brain is presented with new information, it is registered as a stressor. You will learn later that not all stress is bad. During novel mental activity new connections are made between nerve cells in the brain. With time, new branches on the nerve cells are grown (neuroplasticity). Even new brain cells are stimulated to grow (neurogenesis), particularly in brain regions dedicated to encoding and retrieving information (the hippocampal structures). Having the optimal amount and right kind of mental activity, in young and old, is brain enrichment.

When it comes to small animals, it is easy to provide environments that are enriched or deprived of learning opportunities. Those experiments cannot be done on humans, for obvious reasons, but brave little rats and monkeys have sacrificed their lives to show us a great deal about mental enrichment or deprivation.[1]

Mental enrichment: mammalian brain studies

	Enriched environments	Deprived environments
Glial cells	increased	decreased
Brain size	increased	decreased
Dendrites	increased	decreased
Connections	increased	decreased
Size of hippocmpus	increased	decreased
Synapse / neuron ratio	increased	decreased

The above observations deserve some elaboration. They are not the result of a single experiment but a composite and summary of many experiments on non-human mammals. However, empirical data and human population studies are consistent with what animal research has shown us.[2]

Brain regeneration misconceptions

When I was in medical school, not that long ago, we were taught that glial cells were simply a kind of glue that held the brain together (in Greek glia means *glue*). I was happily surprised to learn that research over the last 25 years has shown that glial cells are just as important as the nerve cells themselves when it comes to brain function. The glial cells are cells that are in intimate contact and biochemically coupled with the thinking cells, the neurons. The glial cells, also known as *astrocytes* because of their star-like shape, are cells that feed the neurons. The size and number of glial cells tend to increase upon mental stimulation and activity. The neurons love this. The neurons in response to their own stimulation and feeding increase in size, sprouting new buttons and outgrowths to connect with other neurons. In some regions, notably the hippocampus (HC) and cortex, neurons and glial cells can grow in size and can even multiply, increasing not only in numbers but also in complexity. This takes place even in old age. Unfortunately, even ten years ago, doctors were taught that brain cells did not regenerate or grow. Just a decade ago I met a second-year resident in psychiatry who still believed that. It takes a long time for scientific knowledge to work its way into the trenches of everyday medical practice. Medical science upheld that once early adulthood was reached, brain cells could only diminish in numbers – only a downhill course was possible. We now know that is not true. The brain is very plastic. It responds quickly via neurotransmitters to environmental changes and stimuli and by a slower process, it forms new receptors to accommodate the changes in internal traffic. That process involves protein synthesis which is the formation of new brain tissue!

Another serious misconception is that we associate problems with memory and memory loss in particular, with a loss of brain cells. Some studies on brain atrophy show that many brain cells, even in Alzheimer's disease, are not dead but atrophied and may be reactivated.[3] The connections between neurons are called *synapses* and the result of neuronal activity and challenge is an increase in the number of synapses per neuron. This is what is referred to as the *synaptic density* or *synapse-neuron ratio*. It is important to note that the increase in synaptic density and the increase in the number of neurons represents real growth of tissue – new protein, new life. That explains why there is an increase in brain size, hippocampal size in particular, and brain weight after a period of brain enrichment.[4]

The actual problem in brain atrophy is not just one of cell loss, but a loss of the connections between the cells. In other words there is a degradation or loss of the synaptic density. Changes in synaptic density can be seen in mammals within only a few weeks of increasing or decreasing brain cell activity. The HC can shrink by up to 25% because of lack of stimulation and regain this mass upon reintroduction of brain enrichment.[5]

Studies of adult monkeys in the mid-1960s supported the belief that the supply of neurons is fixed at birth. Hence the surprise, when Elizabeth Gould and Charles Cross of Princeton University reported recently that the monkeys they studied seemed to be minting thousands of new neurons a day in the hippocampus. Even more amazing, Gould and Cross found evidence that a steady stream of the fresh cells may be continually migrating to the cerebral cortex.[6] In the same vein, Fernando Nottebohm of Rockefeller University has been showing that canaries create a new batch of neurons every time they learn a song, and then slough them off when it is time to change tunes.[7]

Brain enrichment

What is brain enrichment? As far as rats are concerned it is merely an interesting environment full of places to explore and things to climb with other friendly little rats as company. Brain deprivation, on the other hand, is characterized by a cage furnished with only water and food. For primates (monkeys and humans), the picture is similar but more complex. An enriched environment for a child may be a home with other people interacting in a friendly way, toys, lessons in music, and so forth. A deprived environment in the extreme could be characterized by what the 'Roumanian orphans' endured. Being left in a crib or bed with minimal human interaction, no toys, no language stimulation and no school. They became mentally retarded for environmental, not genetic reasons. In some cases that was partially reversible as we have found out, because when they were adopted out to caring families, the mental functions improved remarkably but many did not reach expected levels compared to their healthy peers.[8]

We can observe parallel situations with the elderly. In a progressive retirement environment seniors have much autonomy, artistic and social activities and appropriate intellectual challenges.[9] In old fashioned, medical-model facilities, elders are regimented, bored and deprived of chances to participate in life. No pets, no children, just a ghetto-like atmosphere of boredom and routine. Guess which facility type has more agitated behaviours, depression, loneliness and psychotropic drug use? Which is more expensive to operate?[10]

Brain scans (PET scans and fMRI scans) are able to reveal regions of the brain that become 'fired up' in response to certain tasks and activities. The heightened metabolic activity is highly specific to function. PET scans demonstrate that when we hear words, the posterior temporal lobe and acoustic regions are 'fired up'. When we are generating words, the frontal lobes become more active. When we are seeing words, the occipital visual cortex is activated. When we speak words the motor cortex comes into play. When monkeys are gazing at a screen, certain parts of the visual and motor cortex are activated. When they are rewarded, there is a specific response for a correct operation. When they are uncertain or ambivalent, there is a specific response, and when they are making a mistake and are not rewarded, another class of neurons is activated. Those have been nick-named the 'oops' neurons. The regional cortical metabolic activity induced by a particular mental activity or even expectations, releases brain chemicals which facilitate further activity.[11] Some of these effects spill over to other areas because the brain chemicals and hormones generated during these periods do not always stay in one place but circulate. Hence, when schools cut music classes to save money, they found that kids' math scores fell.[12] As part of The Healthy Brain Program, I made a concerted effort to review research in this area and I have found that there is overwhelming scientific evidence that mental activity protects brains.[13] However, it has to be the right kind of activity. Too much activity may even be harmful. Some types of activity promote impulsiveness and violent behavior. Overall, active brains last longer and brain function improves with activity and even to some extent with passive activation. That applies to things like massage, transcutaneous electrical nerve stimulation (TENS), light stimulation, stimulation by sound (music therapy) or smell (aroma therapy). Conversely, deprivation of brain activity leads to brain atrophy as does lack of physical activity to atrophy of muscle tissues. After reviewing the evidence, I can say with certainty that lack of intellectual stimulation is bad for the brain – it causes atrophy – brain drain!

Brain atrophy

It appears from research that the brain needs a certain amount of mental activity and stimulation even to stay at baseline levels. This should not be surprising. I know from personal experience that if I don't exercise for a while my muscles become smaller and weaker. I am sure many people have had this experience. The same thing happens in the brain cells and the same thing happens to their feeder glial cells.

Atrophy refers to a shrinking of structures in the brain or the whole brain itself may shrink. In simple atrophy (the Greek word *trophe* means nourishment; atrophy means without nourishment), brain cells do not suddenly die, they are simply reduced in vitality and size. An organ like the brain may show shrinkage as a result of cell loss (from cell death), cell shrinkage, or a combination of the two. In AD there is cell death but atrophy precedes cell loss. It has been shown that stimulating brain regions with bright light, for instance, reverses some of the atrophy and improves mood and cognitive function. TENS has been used in a similar way.[14] As noted in the previous paragraps, the brain functions in a dedicated modular fashion and it follows that whatever region is stimulated, it will grow by increasing synaptic density.

Studies on brain activity

The great scientist, Albert Einstein, has always fascinated me. He reported that when he was deep in thought his thought process was visual, not verbal. Einstein donated his brain to medical science and it has been studied in great detail. When Einstein's brain was examined and compared to the brains of men who died at the same age (76), and who had normal intelligence, the researchers found that it was not the number of cells that were increased but the number of pathways and connections between the cells. Glial cells were also increased in number and size. That was particularly true in the temporal lobes that are dedicated to language and symbolic thinking.[15]

The London Taxi Drivers Study is also illuminating. Researchers reported that taxi drivers in London had larger hippocampi (one hippocampus on each side) than the general population. That was attributed to their constantly having to use their minds and in particular, those parts of the brain that are dedicated to the encoding and decoding of spatial relationships. When those drivers stopped driving, their hippocampi underwent some reduction in size.[16] That is very analogous to flexing a muscle over time, producing hypertrophy, and upon stopping exercise, the tendency toward atrophy.

Experiments in the study of brain changes resulting from experience (or lack of it) are extremely difficult to conduct in human beings. Even if somebody was willing to volunteer, no ethics committee would ever allow anything even remotely approaching an invasive procedure on the brain. Much of what we know is from animal research or from observational studies and longitudinal studies on living subjects who donated their brains so that scientists could learn from them. There are a number of these famous longitudinal studies that are still running.

One of the most fascinating ones is the Nun Study.[17] Dr. David Snowdon, an epidemiologist, designed a detailed study in collaboration with the School of Sisters of Notre Dame. The school of sisters was founded in 1833 in Bavaria, Germany and they have had a deep interest in providing education and spiritual formation. At the time of the study, about 3200 of the sisters were still ministering in the USA and 678 of the sisters enrolled in the research project. They were from various states but they all had one thing in common: they willed their brains to medical science to be studied postmortem. Not only that, they consented to having their histories scrutinized by scientists and they were followed through their lives by repeated medical and neuropsychological tests. In that way anatomical brain changes found at autopsy could be correlated with lifetime events such as education, intercurrent illness, physical exercise, dietary factors and so forth. What was really unique was that they all had the same lifestyle and constituted a relatively healthy homogeneous population sample with minimal confounding factors that would have been present in the general population (like smoking, drug addiction, poverty, prostitution, AIDS, etc.).

There are other longitudinal studies that I must mention to make the story complete. The Baltimore Longitudinal Study was started in 1955.[18] The Seattle Longitudinal Study was started in 1956.[19] The already-mentioned Nun Study was started in 1986. The Harvard Physician Study was reported in 1991.[20] All of the above studies tend to support the same conclusions. People who have routinized lives who are bored and dissatisfied with life tend to deteriorate. People without deterioration on the other hand enjoy a variety of activities, are open to new ideas and remain flexible. They generally have above average education and income. They have intelligent companionship. They suffer from less chronic disease, less depression and are generally satisfied with personal accomplishments. Of course, this could be a 'chicken and egg' issue. No doubt having good health, good company and money are protective factors. Which comes first, a restricted life style or lack of resources? Animal studies suggest that environmental deprivation, which in human terms is poverty, lack of education, stressful relationships and so forth, translate into compromises in actual brain function. On the other, hand environmental enrichment offers protection against neurodegeneration.

The Harvard Physicians Study of over 1,000 physicians yielded some very interesting insights. Given that this study examined only physicians rules out effects of lack of education, lack of social and mental stimulation and chronic poverty. The doctors studied were divided into two groups: those who used their brains and those who didn't. The 'users' were the ones who, upon retirement, remained active. They continued most of their intellectual activities, they were socially involved and they were productive. They generally enjoyed their activities and found them meaningful. On the other hand, the group designated as 'non-users' became passive after retirement. They withdrew from social life, their lives became simple and repetitive, although they were not unhappy or suffering from any major disease. The results showed that both groups had a steady, slow decline in brain cells, as would be expected with age – generally estimated to be around 500 cells per hour. (Don't forget that there are also some new cells forming, so that is not a net loss.) Brain weight declined accordingly. The synaptic density (synapse-neuron ratio) gradually increased reflecting new learning. However, in the non-users the synaptic density started to decrease and the number of neurons, brain weight and synaptic density all continued to decrease until death. In the users, although the number of neurons continued to decrease, brain weight actually started to rise because of the increasing synaptic density. That was a reflection of both increased protein synthesis and the number of connections and pathways between the nerve cells. This is a difficult concept – think of it as a large forest where single trees are dying at a steady rate. As long as the rest of the trees are healthy and keep branching, the size and density of the forest can actually increase, even as it ages and loses trees. Successful, active aging gives us a bigger and better brain! This is the reason why I get angry when my more pessimistic colleagues remind me that the aging brain is degenerating and losing brain cells. They are correct but they are overlooking the fact that healthy aging is associated with reliable performance and productivity, and that translates into enjoyment of life. Belittling older people on account of their having lost more brain cells than their younger counterparts is a shallow and mean 'blow below the belt'.

In summary, the researches on brain activity found two clear trends associated with aging:

- Between the ages of 20 and 60, reaction time doubles. In other words we slow down. This means that even if we are healthy we can expect to become slower compared to our younger counterparts. We cannot catch a ball as fast and we cannot hit the brakes on a car as fast. No wonder older drivers are bad drivers.
- The synaptic density increases for those who use their brains and decreases for those who do not. Learning results in the formation of new synapses. If we continue to be mentally active, reading books, talking to interesting

people and so forth, our brain connections will actually grow in numbers and complexity. Studies support the happy conclusion that, barring some unfortunate disease, our vocabulary, knowledge and intellectual judgement and performance will increase, not decrease, with age.

- The old brain works hard at making sense of things and is very good at getting the 'big picture' and other types of pattern recognition. Young brains are quicker, often impulsive and better on detail.

Environmental deprivation or even lack of baseline stimulation is a kind of 'brain drain'. High-risk groups are children and the elderly. Animal studies and longitudinal studies on humans have taught me another clear lesson – there are definite protective factors against mental deterioration. These are:

- Social stimulation and social activity
- Multitasking activities
- Active versus passive involvement
- Educational and novel activities

The above protective factors are free and available to all!

Activity, memory and cognitive reserve

As we noted, mental activity should not be excessive, difficult or stressful. The best mental stimulation is pleasurable and somewhat challenging – it should be fun! It neeed not be a contest or a feat. Great minds may produce art, novels, mathematical formulas, but any old brain will benefit from much simpler things – like bingo. Bingo keeps the mind sharp because it involves multitasking and a recruitment of all brain regions. What other activities can be considered mental stimulation and mental activity? There are many schools of thought about mental activities. Depending on what researcher you follow, types of brain activity have been classified in various ways. Broadly speaking, linguistic, musical, math-logic, spatial/kinesthetic, interpersonal and personal activities have been described and have grown to be accepted by the scientific community as edifying forms of mental stimulation. All of the above activities have direct representations on the brain and will involve those brain regions. Some people have thought of them as types of intelligences. While that idea is somewhat controversial there is little doubt that each of the above modalities involves specific memory. In other words we have memory for language, memory for music, memory for mathematical and logical operations, memory for spatial relationships such as our neighbourhood or the location of a street, interpersonal memory pertaining to other people and our relationships with them, personal memory pertaining to ourselves and even memory of having memories (metamemory). Modern society puts a great deal of emphasis on verbal performance to the extent that if a person loses their ability to speak well or does not develop this ability he or she will be erroneously regarded as being stupid or cognitively impaired.

Another way of looking at brain activities is not in terms of the modular location of a particular stimulus, but as trying to understand it in terms of processes. An early pioneer in this area is Edward de Bono who described thought process not with words but with colors.[21] In de Bono's classification, objective thinking was given the color *white*, critical negative thinking was *black*, positive thinking was *yellow*, creative thinking was *green*, intuitive/emotional thinking was *red* and thoughts pertaining to self-monitoring and self-understanding were given the color *blue*. The elegance of this method lies in not having to use a lot of words to characterize types of thinking. It takes for granted that mental processes are largely nonverbal.

Earlier in this book I alluded to protective factors and inner protective mechanisms (neuroprotection, neurogenesis, neuroplasticity, antioxidant activity, etc.) which combat brain degeneration. Neuroscientists call that *brain reserve*, or *cognitive reserve*. Some researchers refer to it as *resilience*, although that term is usually used in a broader context meaning hardiness in old age. Neuroscientists today are very interested in cognitive reserve.[22] There is a growing recognition that diagnosing diseases of brain degeneration in late stages is not very helpful.[23] Trying to define and quantify cognitive reserve is extremely difficult in the study of old folks because the older population is highly heterogeneous. That means there is a huge variation between individuals – no two seem to be alike. When you look at young adults they are generally similar. When you look at old people they are hugely different. A hundred-year-old could be in a fetal position in a nursing home, semi-comatose, or running a corporation using all cognitive skills. Now that we have more and more people surviving into old age, it is actually possible to identify a cohort of high-functioning super-seniors. When we study this population we come up with results that are similar to the already mentioned longitudinal studies.[23] It is clear that some people can remain high functioning well into old age, even above age 100. We know that learning does take place in old age and that logic and vocabulary remain stable. We also know that if NCI develops, cognitive training and mental activity can have a reversing effect.[24] It appears that old brains can and do learn new tricks if they have to, but more slowly. Even old brains can rewire themselves to some degree. If this weren't true there would be no recovery after a stroke, yet we know that quite a bit of recovery can take place.[25] For example, some people can regain their speech and other lost functions. Sadly, if one assumes that regeneration cannot take place, speech and other therapies will not be offered and that will create a self-fulfilling prophecy – that disability after a stroke is untreatable. Another thing we have learned from the study of centenarians is that the brain actively discards unimportant information.[26] This is not exactly the same as forgetting although it is a 'forgetting' in a way. It certainly is not caused by losing brain cells. We know from the study of centenarians that short-term memory and the ability to multitask remain intact but rapid multitasking is more difficult. Another interesting phenomenon exhibited by older brains is the phenomenon of *recruitment*. Older brains function less rapidly and less linearly but are better at pattern recognition and lateral thinking. Older brains are actually more flexible and are more adept and faster in grasping the 'big picture'. For example, an older brain would have less difficulty making sense of the following paragraph than a young brain, because it is more adept at integrating information and does not get hung up on detail. Try reading the following paragraph:

THE PHAONMNEAL PWEOR OF THE HMUAN MNID

Aoccdrnig to a rscheearch at Cmabrigde Uinervtisy, it dseno't mtaetr in what oerdr the ltteres in a word are, the olny iproamtnt tihng is that the frsit and last ltteer be in the rghit pclae. The rset can be a taotl mses and you can still raed it whotuit a pboerlm. This is bcuseaethe huamn mnid deos not raed ervey lteter by istlef, but the word as a wlohe. Azanmig huh? Yaeh and I awlyas tghuhot slpeling was ipmorantt!

A younger brain may find the above tedious until the brain 'catches on'. That takes practice. I have not conducted experiments involving reading materials like the one above, but I have had the opportunity of asking some young adults to read it. Surprisingly they assumed they had dyslexia or in the case of one with, dyslexia, she thought she was just being symptomatic. In any case, with younger brains there seemed to be a greater difficulty recognizing and making sense of the overall pattern. There is little doubt that older brains work differently than young ones but that may not be because of cell loss. The simple notion that loss of recall, especially rapid recall of information is caused by loss of brain cells is incorrect.

Processing of visual information and cognitive function are very much related. Researchers have found that the elderly can keep up with and sometimes outperform their juniors in some mental tests. Multitasking ability, unfortunately, does decline with age. Visual deficits associated with aging are responsible for more than 40,000 bone fractures every year in the USA and it also explains why older drivers are more likely to be involved in accidents at intersections than younger drivers.[27] Researchers are finding that older people can multitask just as well as young people if they are given more time; older people are slower. It has been shown that older people can perform as well on short-term visual memory tests as younger adults but what is most interesting is that they use different parts of the brain. What the researchers are finding is that older brains are recruiting other parts of the brain in order to compensate. In other words, the older brain rewired itself. The latter study showed that older brains are faster than younger ones when it comes to grasping the big picture.[28] In a hockey game it may mean that young people find it easier to follow an individual player while older people are more likely to visualize the flow of the entire game. There has been a radical shift in our view of the aging brain since 1998 when a groundbreaking article appeared in *Nature*. The author was Fred Gage of San Diego's Salk Institute and Peter Eriksson of Sweden's Goteborg University Institute of Clinical Neurosciences. These researchers outlined evidence of new brain cells being formed by the hundreds in the elderly.[29] Old brains are far more flexible or 'plastic' than anyone has ever realized. The social implications are enormous. Focusing on the abilities, rather than disabilities of the aging brain will yield a big dividend.[30] Interestingly, we also know that cognitive performance often does not correlate with the degree of degenerative brain changes. Some people have a lot of cognitive reserve that protects them even if they have actual brain lesions. Perhaps the most famous case of this was Sister Mary, "the gold standard" of the Nun Study.[31] She was a remarkable woman. She had high cognitive test scores before her death at age 101. What was more remarkable is that she maintained that high status despite having abundant neurofibrillary tangles and senile plaques, the classic lesions of AD. In fact she had so many lesions that the researchers who looked at her brain after autopsy could not believe that it was her brain. After double-checking to make sure there were no mistakes, they were amazed to confirm that even a brain with that much pathology and with that many years behind it could function at such a high level. Sister Mary, like many of her high functioning cohort with excellent cognitive reserve, remained active in her old age. Thirty per cent of the subjects who performed well had Alzheimer's lesions. The nuns were not only linguistically active but they had hobbies and pursued activities that required manual dexterity. We seldom think of manual work as mental activity but it certainly is. It may be that medical scientists and physicians are putting too much emphasis on finding lesions and not enough emphasis on brain reserve, how to attain it, keep it and build on it.

Music and the brain

One of the many types of brain activities that deserves special attention is music. There is little doubt that the brain is profoundly affected by sound. It is adversely affected by loud cacophony and beneficially affected by a complex musical relationships with the rhythm approaching that of the heartbeat. It has been noted some time ago that certain types of music can reduce the frequency of seizures in epilepsy. Music has been used in relaxation therapy. In particular, some of Mozart's music is effective in this manner. That has been called the *Mozart Effect*. His sonata K448 for two pianos has been used to reduce anxiety and to treat epilepsy.[32] This type of music has been known to boost spatial reasoning powers after ten minutes of exposure. On the other hand, minimalist music or just repetitive noise is not associated with any benefit. Similarly, it has been noted by researchers that children who have had piano lessons for six months had increased spatial reasoning and mathematical skills.[33] Computer games and noise will not have this effect although they too stimulate the brain. In addition to the Mozart sonata, music effective in producing the above phenomenon include some of Bach's and Yanni's compositions. What seems to count is the mathematical complexity and long-term periodicity. Perhaps this is why minimalist and pop music has no such effect.

Approaches to mental fitness

New findings along with older research evidence support the concept that mental exercise may help in training brains to perform better. The nick-name of *neurobics* was born. Can neurobics do for the brain what aerobics can do for the lungs and cardiovascular system?

I have often wondered whether mental fitness could be developed just like physical fitness. Can there be such a thing as a 'mind gym'? A New York City neuropsychologist, Dr. Mark Goldberg certainly believes that.[34] When you go to Dr. Goldberg, he will give you an assessment of various cognitive functions much like a personal trainer would assess your ability for physical strength, flexibility, endurance and balance. The physical trainer would then draw up a plan to improve your fitness based on your strengths and deficits. Similarly, in Dr. Goldberg's mind gym concept one would be prescribed mental activities to activate function and restore balance. By this process of selective activation, a 'brain exercise circuit' is achieved. Dr. Goldberg's method is based on his research on what builds brain reserve. Cognitive fitness is the best protection one can have and it is achieved by enrichment, variety, challenge without much stress and without tedious routine. New learning, multitasking, planning, decision making, manual activity and improvement of manual dexterity are key.

There are many studies that show the benefits of various types of mental activity but this is a young science and its applications lack testing and systematization. How does it help to know that sequencing and categorization tasks can be used to improve executive function? How do we apply findings that support multimodal training involving executive functioning and other more general cognitive tasks, to produce generalization of positive training effects? Fortunately, memory training works best with a multifactorial approach which include encoding activities, subject characteristics, retrieval factors and the nature of the materials – a complex environment seems best. Relaxation procedures improve memory performance because stress interferes with learning and performance. (More about this in Chapter 12.) Memory performance improves when people carry out elaborate encoding procedures by reflecting on information and relating it to things they already know. This means that we have some conscious control over how efficiently we learn. We are not merely sitting ducks in a memory arcade. What is even more intriguing is the finding that our beliefs about memory influence its efficiency. Unfortunately, beliefs about memory become more negative and nihilistic as people age. Beliefs, however, do not positively correlate with performance on artificial memory tests. Overall, much scientific support exists for the effectiveness of several forms of cognitive rehabilitation for persons with cognitive problems, from stroke and brain injuries. I strongly maintain that neurobics and other types of mental stimulation have an important place in the building of brain reserve for the prevention, treatmemt and rehabilitation of damaged brains.

The young brain

So much for old brains. What about baby brains? Over the past few years, an entire industry has arisen around the belief that an enriched environment promotes the development of babies' brains. Although some of the claims are certainly exaggerated, it is well known from animal studies that an enriched environment can promote changes in both behaviour and brain structure.[35] How this happens is still unclear, but a groundbreaking study by Dr. Joe Tsien and colleagues at Princeton University used a combination of genetics, behaviour and electron microscopy to look at the possible basis for this effect.[36] Tsien, whose study on 'intelligent' mice made international headlines, had previously developed a sophisticated genetic trick to knock out genes in a specific part of the brain in the HC, which is central to learning and memory. He had used this method to study the function of a neurotransmitter receptor called the NMDA receptor, which plays a major role in the synaptic changes thought to underlie many types of memory. Mice whose HC lacks NMDA receptors are known to have difficulty finding their way through a maze, and the authors show that they also have poor memory for non-spatial stimuli such as smells and unfamiliar objects. Is this defect genetically hardwired, or can it be overcome given the right environment? Tsien and colleagues tested this by allowing their mutant mice to live in what amounted to a rodent playpen, filled with toys that were changed every few days to stimulate their exploratory tendencies. After two months of exposure to this environment, the mutant mice had overcome their deficits and performed as well as their normal littermates on almost every memory task in which they were tested. The results suggest that something happens in the brain that allows it to learn even without NMDA receptors i.e. even with a major genetic defect. To find out what the defect was, the authors used electron microscopy to look at the fine structure of the HC. They found that the enriched environment led to the formation of more dendritic spines, the tiny structures on which synapses are formed. That signified a denser network of neural connections, a higher synaptic density, and it is tempting to think that the increased density somehow allowed the mice to learn even in the face of a genetic defect that would otherwise prevent learning. That may be hard to prove, as it is difficult to be sure that changes in the HC, rather than elsewhere in the brain, were responsible for the improved performance. Whatever the answer, the results make it clear that memory deficits need not be fixed but can be overcome given the right environment. As with any such study, one must be cautious about extrapolation to humans. Nevertheless, the 'zero-to-three' lobby might take note that the effects of environmental enrichment are not confined to early development. The mice were almost 2 months old before they were transferred to the enriched cages, which in human terms would correspond to early adulthood.

There is a big world out there full of enrichment opportunities. That should not blind us to what we know about deleterious effects. What kind of stimulation is a bad for the brain? We know that sheer information overload is bad for the brain because it provokes a stress response accompanied by elevation of stress hormones that are specifically harmful to certain brain regions. (More about this when I talk about stress in Chapter 12.) Generally, a repetitive, rapid action kind of stimulation, such as seen on television a great deal of the time and in many video games, produces impulsiveness and a tendency to violence in young people (particularly males).[37] If your TV makes you feel violent please go ahead and kill your TV! Too little intellectual challenge and stimulation is a type of sensory deprivation associated with brain cell atrophy, not necessarily brain cell death. Of course, forced involuntary mental activity is registered by the brain as a stressor, and the concomitant rise in stress hormone levels can actually damage the brain. The way to brain health, for young as well as old brains, is variety, novelty, autonomy and the enjoyment of mental activities.

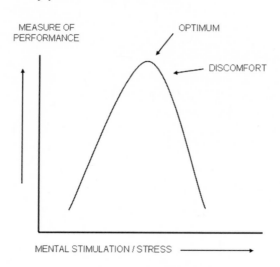

RELATIONSHIP BETWEEN
PERFORMANCE and STRESS

Food for thought

During most of this century, education was for the young. Further learning was not considered possible, or advisable for adults. Since the 1950s, however, scholars have come to believe that adults pass through different developmental stages, and learning continues throughout the lifespan. Given that we have about 30 years more life expectancy than people 100 years ago, it behooves us to continue growing and developing, and to make education a recurrent lifelong activity.

Forgetting

I have noticed that people are almost obsessed with memory and the ability to recall information – often trivial information. Everyone seems anxious about loss of memory these days. People who never worried if they did not recall a name or misplaced a key, upon turning middle aged, become mortified at what they almost always misinterpret as AD. It never occurs to those people that they always remember that they forgot! They have their metamemory. Metamemory, the tracking of mental events and executive function, is more important than memory for facts and places. It makes more sense to be concerned about the executive functions than the ability to recall lists.

Most of us think of 'forgetting', or the loss of memories, as some sort of loss associated with loss of brain cells or brain function. This is not always the case. Neuroscientists have discovered a molecule called cyclic AMP response element-binding protein (CREB), which appears to hold the balance between remembering and forgetting.[38] CREB is a 'transcription factor'. That means it is responsible for switching on particular genes, thus allowing the manufacture of the proteins for which those genes are responsible. It comes in two forms; the *activator CREB*, and the *blocker CREB*. Elevated levels of the activator are found during formation of a memory. These activated genes are important in forming long-term memories. Animals deficient in activator CREB can only learn on a short-term basis, and are unable to recall after an hour or so. The blocker CREB is a slightly different molecule, which interferes with the action of the activator CREB. It prevents the formation of memories. If we did not have blocker CREB, the result would be an over-proliferation of memory.

I thank my mom for the good memories

Anecdotal and correlational evidence abounds that early parental neglect can have severe consequences for children's later cognitive development.[39] Under normal conditions, how does maternal care contribute to brain development and what are the mechanisms? Michael Meaney and colleagues at McGill University provided some interesting clues by studying the offspring of two different groups of female rats: a) those that were well above average in various aspects of maternal care (licking, grooming and arched-back nursing of their pups) and b) those well below average. The major result was that the pups from the high-care mothers performed better in a test of spatial learning and memory. The effects were already apparent at an early age and were still significant in old animals. Because such results were known from other studies to depend heavily on activity in the HC, the authors then compared the HC from the two groups of animals. They found that those pups from high-care mothers also had more synapses in their HC, along with a higher levels of inputs from other brain regions involved in learning. By comparing levels of trophic factors and neurotransmitter receptors, the authors provided evidence that the mechanism that mediates the effects of maternal care might start with a neurotransmitter receptor already known to be involved in learning and plasticity – the NMDA receptor. One other possibility is that the effects could be genetic – i.e. the high-care mothers might have different genes from low-care mothers, which could explain the difference between their offspring. However, that cannot be the whole explanation. The authors performed cross-fostering experiments, and showed that pups born to low-care mothers and raised by high-care mothers were indistinguishable both in performance on the learning and memory tests and brain development. In other words, pups that would normally have performed poorly on the tests did better as a result of receiving better maternal care. Interestingly, the converse was not always true. Pups born to high-care mothers performed well even when transferred to foster mothers who gave them less care. Thus, maternal care influences brain development, but some pups are more sensitive to that influence than others. The authors suggest that the difference may be due to differences in brain chemistry that are already established by the time of birth.[40]

Much more research is needed to understand just how maternal behaviour is translated into changes in brain development and why it affects some individuals more than others. Obviously, great care needs to be taken in generalizing results from animal experiments to humans. Nevertheless, population studies suggest the same positve trend in correlating sustained higher levels of mental functioning throughout the lifetime with previous experiences of enrichment. The present results point to new directions in studying a question that is of fundamental human interest. The science of *epigenetics* explores how environmental impacts (experience), acting through the brain, shape the expression of genes and, in effect, what we become. Be careful about the quality of experience and information that goes into your brain!

THE LEAST YOU NEED TO KNOW

- Mental enrichment is vital for all ages.
- Mental activity 'lights up', feeds and stimulates brain.
- Glial cells nourish thinking neurons.
- Neural connections multiply and grow with age.
- Use it or lose it! – like muscle strength.
- We live in mentally lazy society (e.g. TV).

START NOW!

- Read books, enjoy art, music, dance.
- Engage in some social activities.
- Get involved.
- Educate, teach.
- Reduce stress to optimize learning – enjoy.
- Beware of passive entertainments.
- Avoid overload and noxious stimulation.
- Seek enriching environments for your children, self and elders.

www.HealthyBrain.org

CHAPTER 11

SLEEP

Laugh and the world laughs with you.
Snore and you sleep alone.

– Unknown

Sleep and dreaming have always been a fundamental part of human, if not all mammalian, existence. Many early writings on these subjects were almost entirely speculation and superstition. Long, long before sleep was understood, people have felt the need to explain this amazing phenomenon. Greco-Roman concepts of sleep were based on their belief that there were gods and goddesses that controlled the minor and major events of their lives. They identified the goddess of night, *Nyx*, who had two sons: *Hypnos*, the God of sleep and his brother *Thanatos*, the God of death. Hypnos sprinkled drops of poppy milk into people's eyes so that the opium would make them fall asleep and then fanned sleeping persons with his wings to enable them to sleep in comfort. The great Greek poet Homer wrote that sleep had a "thousand children", the *Dreams*, who hid in a cave in the mountains. The river of *Lethe*, or river of forgetfulness flowed through this cave.

In ancient Greece, if a citizen was unable to sleep because of some problems, he or she visited a sanatorium, one of many dedicated to Asclepius, the Greek healer who became the demi-god of medicine. Here the afflicted spent three weeks in rest, thought and education, soothed by gentle music and then having his or her balance restored, would be able to sleep again. Obviously this was before third party or government run health care schemes. The symbol of Asclepius, a snake coiled about a staff persists to this day as the symbol of the medical profession.

From the time of the Middle Ages until the Renaissance, there were discrete changes in the concept of sleep. More concrete explanations of sleep were enunciated by Lucretius, the Epicurean poet and philosopher, when he described sleep "as an absence of wakefulness". That remained the prevailing view for centuries. In a famous book titled *The Philosophy of Sleep*, published in 1834 by Robert Macnish, member of the faculty of physicians and surgeons of Glasgow

School of Medicine (state-of-the-art at the time), it was put forth that sleep was an intermediate state between wakefulness and death. In other words, until fairly recently, it was believed that when you were asleep you were half dead. Sleep was considered to be a passive state. These early notions gave way to more modern concepts of sleep.

During the 20th century scientific observation and experimentation abounded. As with other brain research, almost everything we know about sleep has come to light in the last few decades. With the discovery of rapid eye movement (REM) sleep we came to realize that during sleep the brain was very much alive and active. Still, it was also believed that although the brain was active, it was struggling to gradually dissipate toxins and poisons, built-up during the day, before the brain could wake up. In the 1920s at the University of Chicago, the first true sleep researcher, Nathaniel Kleitman, contested the idea that sleep was a result of a buildup of toxic substances. He noticed that when people did not sleep the whole night they were most tired during the middle of the night and not the next morning. If toxins were gradually building up during the wakeful state, such recovery without sleep would not be feasible. Such brilliant observations have led to his breakthrough concept that it was not sleep that needed to be explained but wakefulness. This led to some theories that, indeed, there must be different kinds of wakefulness and different kinds of consciousness during various stages of development and types of activity. The new way of looking at sleep was that it was a letdown of the day's highly attuned waking activity.

In 1875 the Scottish physiologist Richard Caton first demonstrated electrical activity in the brains of animals. It wasn't until 1928 that Hans Berger recorded electrical activity in the human brain. The *electroencephalogram* (EEG) was born. Berger also demonstrated clear differences between sleep and awake rhythms. This was the first time in history that brain waves were measured and recorded without disturbing the sleeper and, equally important, that sleep could be measured continuously and quantitatively throughout the whole sleep cycle. Other ingenious experiments, including ones where whole brains of animals were isolated, showed that external stimulation was absolutely necessary to maintain brain activity and wakefulness.

Sleep Disorders

Through early observations (well over 100 years ago) two sleep disorders were clearly defined. One was *narcolepsy* or a sudden sleep attack. The word 'narcolepsy' comes from the Greek word *narcosis*, meaning 'benumbing' and *lepsis*, meaning 'to overtake'. The victim of narcolepsy is suddenly overtaken by an irresistible urge to fall asleep. This is a dangerous condition. The attack is often triggered by laughter, surprise or some other strong emotion or even driving.

Narcolepsy is a discrete syndrome, far more profound than daytime sleepiness experienced by all of us at one time or another. Fortunately, this is a rare condition affecting less than two people in 100,000.

The other syndrome noted was what is today called *obstructive sleep apnea* (OSA).

Sleep apnea in antiquity was described in the fourth century, BCE. The physicians of Dionysius, who had sleep apnea, prescribed that he should get some fine needles, exceedingly long, to thrust through his ribs and belly whenever he happened to fall into a deep sleep – then he would be thoroughly aroused. Magus, King of Cyrene (258 BC) was weighted down with monstrous masses of flesh in his last days. In fact he suffocated much like the victims of OSA who experience choking during the night. The syndrome was described by the novelist Charles Dickens. In 1836 Dickens published a series entitled the *Posthumous Papers of the Pickwick Club*, in which he described Joe, a loud snorer who was obese and somnolent. Joe had difficulty breathing during the night and staying awake during the day. He had some heart trouble too. In any case, that was the first description of the OSA syndrome. It was later written up in the medical journal, *Lancet* in 1877. If you are a respirologist you should have a picture of *The Pickwick Papers: Joe, The Fat Boy* by the artist Joseph Clayton Clarke. The print is popular and is available over the Internet. The topic is fascinating because the term 'Pickwickian Syndrome', although recognized since 1877, was not coined until 1956, by Dr. Burwell who associated it with loss of oxygenation in the lungs and consequently in the blood (and brain). Of course we do not really know what was wrong with Joe. Historically, medical people have hypothesized that he had a lot of other problems besides obesity induced OSA.

Not all sleep apnea is the result of OSA. There are other causes of sleep apnea besides obstruction of the airway. By definition, sleep apnea is characterized by at least 10 seconds of cessation of breathing. I am not trying to make you into a sleep expert but OSA is so common that you should know something about it.

The documented record for the cessation of breathing, the diagnostic feature of sleep apnea, which is really choking, is well over two minutes. (Try holding your breath for that long.) We do not and never will know what the limits are to this kind of asphyxia because sleep laboratories do not allow the apnea to progress. Like Dionysus, their patients will be awakened before they run the risk of death or damage from lack of oxygen.

The prevalence of sleep apnea and OSA

	Men	Women
Sleep apnea	24.0 %	9.0 %
Daytime sleepiness	15.5 %	22.6 %
OSA	4.0 %	2.0 %

Loud snoring is the most easily identified sign of OSA but not snoring is common and does not always signify lack of OSA.

The snoring continuum

- Simple Snoring
- Upper Airway Resistance Syndrome
- Obtructive Hypopneas
- Mild Obstructive Sleep Apnea
- Moderate Obstructive Sleep Apnea
- Severe Obstructive Sleep Apnea

People who sleep alone cannot be aware of how they sleep. Spouses or bedfellows of those with OSA, however, are acutely aware of their loud snoring. They are often the ones who bring the sufferer to medical attention because of their own sleep deprivation. They report loud snoring with pauses in breathing, gasping and choking, restless sleep and sleepiness in the morning. The victim of OSA feels groggy and tired in the morning and even during the day, but usually improves after the morning as he or she gets more oxygen. He (typically it is a male) often feels irritability, depression, sexual dysfunction, headaches, memory loss and lack of concentration. There is a clear association with high blood pressure, heart disease, stroke, accidents, poor job performance and job loss. All of the following conditions are made worse by OSA: diabetes, hypertension, cardiovascular disease and cardiac arrhythmias.[1] It is also correlated with stroke and motor vehicle accidents.[2] Sleep apnea can also kill.

Common sleep disorders

Since the early description of these syndromes, modern medicine has defined many types of sleep disorders. The following is a list of some of the more common ones which may come to your doctor's attention:

- Insomnia (trouble falling asleep, maintaining sleep or not getting restorative sleep)
- Obstructive sleep apnea (OSA)
- Narcolepsy
- Restless leg syndrome (RLS)
- Sleep paralysis
- Hypnagogic hallucinations
- Rapid eye movement-associated behaviour disorder (REMBD)
- Somnambulism
- Night terrors

Fascination with dreams leads to sleep science

Back to a little more history. By far the most widespread, although indirect, interest in sleep by physicians was engendered by the theories of Sigmund Freud. His interest was really in dreaming and not in the physiology of sleep. Freud developed psychoanalysis and the technique of dream interpretation as part of his therapeutic approach to emotional and 'mental' problems.

It was believed that dreams were the guardians of sleep and they occurred in response to a mental disturbance (i.e. anxiety) to avoid waking up. Freud had a theory that dreams discharged instinctual energy and this gave rise to a notion that dreaming was a kind of safety valve for the mind. In fact Freud's interest in dreaming was so influential that dream research became far more important than sleep research. Academic psychiatrists and psychologists and indeed people all over, especially in North America, were avidly interpreting one another's dreams. Sleep research gained its initial popularity and financial support as a result of an almost obsessive pursuit of meanings from dream interpretation. Ironically, as belief in psychoanalysis was reaching almost religious proportions and dream analysis likewise, fundamental scientific sleep research was able to hitch its wagon, so to speak, onto this ardent pseudoscientific movement. Following that, much of the medical and psychiatric establishment started to take dreaming more seriously. Ultimately it was not important which came first, the proverbial 'egg' of dreams or the 'chicken' of sleep research. It all ended up as the same circular and fascinating exploration into the importance of dreaming and non-dreaming sleep.

With the observation that sleep cycles existed and that there were dreaming and non-dreaming phases to sleep, the fascinating world of *chronobiology* opened up. Most sleep specialists now share the opinion that chronobiology, the study of biological rhythms, is a legitimate part of sleep research and indeed all endocrinology research. The 24-hour rhythms in the activities of plants and animals have been recognized for centuries. Those biologic rhythms were reasonably assumed to be a direct consequence of the 24-hour environmental fluctuations in light and darkness. Further research into chronobiology discovered that internal clocks exist in animals and humans that operate independently of the day-night cycle. It appears that the sleep onset is in response to daylight cycles but dream and non-dream phases of sleep respond to some inner clockwork independent of fluctuations in daylight.

The body is ready to sleep when the core temperature drops. That event is mediated by the hormone of darkness, melatonin (also see Chapter 13 on hormones). Melatonin does not act like the usual sleeping pill; its effect is subtle – it is not a sedative drug – it does not depress the central nervous system the way tranquilizers do. It merely resets the inner clock tricking the brain into feeling as though it is dusk. You can achieve the same thing by exposing yourself to bright light during the early morning hours. That advances the body clock. Adolescents have a clock setting that produces a delayed sleep phase. Their core body temperature does not drop till well after midnight. That explains why teens tend to have trouble going to sleep at the same time as their parents and why they are tired in the morning. They are not lazy. They are not just rebellious nor are they ill. They have a normal physiological state – delayed sleep phase when compared to the conventional adult world. You had this too when you were a teenager. By exposing teens to bright light in the morning, their clocks can be reset so that they will feel sleepy earlier.

For the aged the reverse is true. Old people experience a drop in core body temperature at an earlier hour – around 7:00 p.m. They experience early morning awakening, say 4:00 a.m., and are often alarmed by that, wishing they slept longer – to be in synchrony with the conventional world. The solution for these persons is to use bright lights in the evening, which pushes their biological clock back and the core body temperature will drop later. The bright light method is a particularly clever and safe intervention because elderly people are very sensitive to sedative drugs which usually do them more harm than good.

It is important to note that in each of the above phase shift problems, the role of the bright light is crucial. Merely trying to get into the habit of getting to bed early (for teens) or staying up later (for elders) will not work because the core body temperature will not align and the body will not be ready for sleep. The day/night cycle of the adult working world is assumed to be the norm and desirable. That is merely a convention to enable adults to work and earn a living – it does not serve the young and the old very well.

It is now known that there is a discrete phenomenon known as *rapid eye movement* (REM) sleep. What is interesting about REM sleep is that if you awaken persons during the REM phase, they will report visual imagery and physical activity. A great deal of tedious research went into the observation of infants and adults during sleep culminating in night-long sleep studies. Information was also correlated with the EEG and reports of awakenings from REM sleep. It became clear that during REM periods subjects were actually hallucinating and motor systems were short-circuited, deep in the brain, to shut off physical action responses. In effect, during REM sleep when all this hallucinatory activity was going on, we are paralyzed. Detailed EEG tracings showed that REM sleep periods occurred several times during the night. The discovery of various EEG patterns during sleep led to the discovery of what was nick-named *sleep architecture*. When we look at the sleep architecture, interesting patterns emerge. We have 90-minute cycles of non-REM sleep interspersed with 4-5 REM periods. Non-REM tracings were correlated with a deeper sleep. That deep sleep, characterized by slow waves on the EEG (also called slow-wave or SW sleep), is of a very different quality and physiological significance. All that was quite mysterious when it was being discovered and to a large extent it still is!

When we are asleep there is a dual process going on. There are dreaming periods (REM sleep) and definite non-dreaming periods (the non-REM cycles). REM sleep is qualitatively very different from non-REM sleep. While there is a great deal of visual and inhibited motor activity going on during REM sleep, this is not true for non-REM sleep. It is a significant feature during REM sleep that the brain is much more active and it is certainly not in deep sleep. We are more easily aroused and in fact dreams themselves are sometimes arousing and will provoke an awakening. On the other hand, during deep sleep arousal is more difficult and entirely different physiological processes are at work. During deep sleep the brain appears to be more at rest and more fundamental physiological repair processes are activated.

It will be instructive to look at the neurology of wakefulness and sleep in more detail. Different brain waves and brain regions are activated during the different phases of consciousness and, not surprisingly, there are different important physiological processes initiated.

We tend to be either awake, in deep sleep or asleep and in a dream. Those states are characterized by specific changes in the EEG. During stages of sleep and wakefulness the EEG brain waves assume certain characteristics. Wakefulness is characterized by well synchronized regular patterns of what are called alpha waves (8-12 Hz). The *alpha state* occurs when individuals are resting with their eyes closed but awake. Try it for a few minutes. Close your eyes and sit calmly. You will be experiencing alpha waves or the alpha state. The alpha rhythm is blocked by opening the eyes or paying attention to something or doing some mental work. That phenomenon is called *alpha blocking*. As you read this your alpha waves are blocked. Transition to sleep is not an all-or-none phenomenon but a continuum. Boredom and fatigue is characterized by diminishing alpha waves and the appearance of slower *theta waves* (4-8 Hz). That is called *alpha dropout*. I hope you are not experiencing this right now. As the alpha dropout increases, drowsiness increases, the EEG becomes somewhat jumbled and desynchronized and strange waves called *sleep spindles* appear (12-14 Hz). Sleep spindles are the hallmark of the onset of true sleep where stimuli from the outside world are shut off from the cortex. (Remember, the cortex is where consciousness resides.) *Deep sleep* or SW sleep soon follows. SW sleep, as the name suggests, is characterized by strong, very slow waves (0.5-3 Hz) and during deep sleep there is no eye movement or any other activity. The brain is truly asleep in the sense that it is at rest, off-line as it were, and is allowing other processes to kick in. It is during this deep restful sleep, without dream activation, that we experience a boost to the immune system and other hormonal systems to promote cell differentiation and growth – it is a time of anabolic activity and repair. Growth hormone is secreted and growth (in children) takes place during deep sleep.[3] The immune system goes to work reducing infection and attacking cancer cells. People who are sleep deprived have increased rates of infections and cancer.[4] Deep sleep is periodically interrupted by REM sleep. REM sleep has also been called 'paradoxical sleep' because it appears to be sleep but as we have seen in the previous paragraphs, it is a time of much brain activity. During REM sleep the brain is blocked from activating the body musculature but it is far from being at rest (as in deep sleep). The REM sleep periods increase so that as sleep progresses throughout the night they become more and more dominant until we awaken, often at the tail end of a dream – the only one we tend to recall.

Definition of healthy sleep: You fall asleep in about 15-20 minutes, awakening only if necessary (to void for example) and you fall back asleep easily, awakening spontaneously, feeling refreshed and alert in the morning.

Comparison of dreaming (REM) sleep and deep (non-REM) sleep

Measurement	REM sleep	Non-REM sleep
EEG mixed frequency	Low voltage, slow waves and spindles	Higher voltage,
Eye movements	Conjugate rapid	None or few slow movements
Muscle tension	Almost absent	Present, less than in wakefulness
Body movements	Twitching	Few gross movements
Respiration	Variable, shallow	Regular and deep

Heart rate	Variable, rapid	Regular and slower
Blood pressure	Variable	Below waking level
Penile erection	Present in dream sleep	Absent
Mental activity	Dramatic, dream-like	Repetitive, thought-like

Recall that we alluded to various processes inside the brain that coincided with wakefulness, sleep and dreaming. During wakefulness, the Ventral System (so called because if the brain was flattened, it would be on the stomach side) is firing using acetylcholine, serotonin, dopamine and norepinephrine, while on the top, the Dorsal System, is busy firing with acetylcholine as the main neurotransmitter. In deep sleep, inhibitory signals mediated by (gamma-amino-butyric acid) GABA neurotransmitter, act on cortical, limbic and motor centers. In REM sleep, cortical areas are activated but desynchronized. The neurotransmitters GABA and glycine inhibit motor neurons to produce muscle paralysis.[5]

In 1949, Horace Magoun of Chicago and his students identified the *ascending reticular activating system* (RAS) – direct stimulation of which activates or desynchronizes the EEG, replacing high-voltage slow waves with low-voltage fast activity. Note that the RAS gets signals from the outside world and inner organs and sends signals to the cortex, promoting wakefulness.[6] Note that in deep sleep, specific brain regions from the brainstem act on the excitatory regions to shut down stimulation to the cortex. During dreaming specific regions act to stimulate the RAS to fire up the cortex but these signals are not coming from internal organs nor the outside world via the spinal cord – the brain is busy reorganizing itself.

Neurotransmitters levels affecting sleep/wake cycles

Neurotransmitter	Sleep/wake state Wakefulness	Deep Sleep	REM Sleep
Acetylcholine	high	low to none	high
Serotonin	high	low	none
Norepinephrine	high	low	none
Histamine	high	low	none
Orexin	high	low	low
GABA	low	medium	high

The terms *sleep dependent consolidation* or *sleep dependent plasticity* refer to a phenomenaon well known to sleep researchers, teachers and coaches. Simply, it means that after an initial training or learning period, sleep is necessary for further brain rehearsing and learning, which occurs during a good night's sleep.

Students and athletes deprived of regular sleep between training/learning sessions do not perform optimally.[7]

A good night's sleep

By now you have the idea that sleep is complex and important for optimal performance on many levels. That is actually an understatement. Sleep is vital. It is so important that mammalian life hardly exists without it. The average person will become delirious and psychotic after missing just a few nights sleep. In some vulnerable individuals, particularly children, the aged, those already tired or just constitutionally predisposed, even one night without sleep can precipitate gross disruption of functions. There is a progression of events caused by severe sleep deprivation that we can only glean from accounts of torture victims because such experiments are not allowed in medicine. After severe fatigue and mental derangement the body goes into a state of irreversible hypothermia and the victim dies. That usually happens after only six days. Yes, sleep is indeed vital!

In ordinary civilian life we do not see such extremes but milder forms of sleep deprivation are rampant, with terrible consequences.[8]

Sleep deprivation-related events[9]

- 25,000 deaths per year related to the impairment of the immune system
- 25,000 accident deaths (not counting automobile accidents)
- US National Highway Traffic Safety Administration links 100,000 crashes a year to drowsiness
- 2.5 million disabling injuries from the above accidents (after the doctors are finished, this can keep lawyers busy for years at a time)
- Cost to the taxpayer: $ 56 billion/year in the USA. (For Canada, divide by ten)

We are a sleep-deprived society. We put high value on productivity and sleep is not productive – or so it seems. But the price of that illusion is very high. Major disasters have been associated with workers being 'asleep at the switch'. Those include the Exxon Valdez oil tanker spill, the Chernobyl nuclear plant meltdown, the Bhopal chemical plant contamination and the Three Mile Island nuclear disaster. It is also known from studies on physicians that long working hours and sleep debt adversely affect physician health and patient safety.[10] The cost in dollars is staggering; the cost in human suffering is unimaginable. Recall that prior to 1913, before the light bulb, people slept an average of 9 hours per night – 20% more than what people get today. Studies show that the average person needs 8.15 hours of sleep every night. This is an average – some need more, some need less.

The average North American only gets 7.5 hours per night and many people get less than that.[11] How can you tell how much sleep you need? It turns out that the amount of sleep we need is fairly constant throughout our lifespan. As we age we do need a little more sleep. Contrary to popular belief, studies show that children, the elderly and the sick need more than the average amount of sleep.[12] If you look at times when you were healthy and happy and can estimate how much sleep you got, that may be a good indicator. (Ask your mother!) There are also some scientifically validated assessment tools for estimating sleep debt. Read on and you will learn how to figure out how much sleep you really need. Just remember that you can't cheat on sleep. Your brain will extract what sleep it needs at any cost to you, your loved ones and your co-workers. When your brain needs sleep it will put you and the public at risk. The situation is worse than you thought. Sleepy people are dangerous!

Sleep is becoming a commodity – Sleep for Sale

Tired New Yorkers in need of a power nap could get it at Metro-Naps for $14 for 20 minutes. Metro-Naps was located in the Empire State Building. Company co-founder: Christopher Lindholst.

What about naps? In normal people, performance drops off sharply in the afternoon, and we do mean after noon – just after lunch. It is easily restored to morning levels by a short (say half-hour) nap.[13] The trick is to get a short nap without bedding down and getting too much sleep. That would, of course, throw off the biological clock and set the whole sleep-wake cycle topsy-turvy. Ironically, because of cultural norms, North Americans and Europeans do not get enough sleep while our southern counterparts, a full two-thirds of the world enjoy afternoon naps (i.e. the *siesta*). The natural pattern is to nap. If you don't believe this check out the *Polar Psychology Project*.[14] Most of the world has a nap! The only caveat is that it not lead to day-night reversal because that is a true illness – a circadian rhythm disorder. When it comes to normal sleep, like in other things, there is some variability. There are differences according to gender and age.

Let's look at gender differences first. Men enjoy less deep sleep (SW sleep) than women. Men have less natural growth hormone, secreted during deep sleep, than women. (Don't jump to conclusions – that is not why women live longer.) Sleep disorders are also common during menopause.[15]

Sleep structure and patterns also change with age. Children need intact deep sleep for proper growth and emotional development. Behavioral problems develop when bedtime is irregular or too late.[15.1] The elderly generally need a little more, not less, sleep. The popular illusion is that older people need less sleep. What is true is that older folk get less sleep because they have more sleep disruption, usually from chronic physical ailments (e.g. bladder problems). As we age, sleep efficiency declines. That means we spend more time in bed but get less refreshing sleep. That is not a result of healthy aging but a complication of aging. The result is insomnia of the medical kind. The amount of time in restorative, slow wave sleep declines as we have more awakenings. It has been said that "getting old is not for sissies". I can also add that "getting old is not for those who enjoy sleep". That is truly tragic! For older individuals, sleep stages 1 and 2 get progressively longer while stages 3 and 4 (SW sleep) shorten. This implies less time in deep, restorative slumber and more in light sleep. Sleep efficiency decreases, and the time taken to fall asleep (sleep latency) increases. A phase shift occurs towards the earlier hours dictating earlier bedtime and earlier awakening in the morning. In a study of more than 9,000 patients over age 65, only 12% reported no sleep complaints. More than 50% reported chronic sleep problems.[16]

There is more bad news. Generally, about 50% of men over 50 years old suffer with elevated stress hormone levels in the evening and that state is associated with fragmentation of sleep. Older men also have less deep sleep than women, and as noted previously, deep sleep deprivation is associated with lowered growth hormone levels.[17] There is little doubt that the endocrine (hormonal) system and its regulation are very important to maintain healthy sleep patterns. (For a more thorough discussion on the effects of hormones and the role of growth hormone wait till you get to Chapter 12 on the role of stress and Chapter 13 on hormones and the brain.)

Alcohol and sleep loss

Sleep loss may be worse than alcohol for impairment. Chronically sleep-deprived people score as badly or worse on reaction time tests than people who are legally intoxicated (0.08% alcohol).[18] Several 18-hour days leave you with impairment. A 24-hour day without sleep is impairment. Any alcohol is additive to that. Thus, an 18-hour day and a drink = two drinks, and so forth. You may be driving with a low (less than 0.8%) alcohol level and still be impaired. Beware!

Insomnia

The next topic is more mundane but actually more important – the common condition generically known as insomnia. Most of us manage to avoid accidents and do not cause disasters despite our sleep debt, which we all do pay back at one time or another. Insomnia is so common that it is safe to say that everyone has experienced it, at least in its milder forms. One night of poor sleep every week or so is almost universal. Medical insomnia, a real clinical syndrome, is more severe and more serious – it requires diagnosis and medical treatment. By definition, *primary insomnia* exists if there is difficulty initiating sleep or maintaining sleep, or not getting restorative sleep for at least one month causing some impairment of functioning and/or distress. It is part of the definition that it not be caused by another of the sleep disorders (as described below), a medical illness, psychiatric illness or some drug or substance use.

Facts about insomnia and sleep disorders[19]

- About 10-20% of adults report poor sleep up to three times per week.
- Depending on what study you read, the prevalence of chronic insomnia is 9-18 % of the adult population.
- Among the elderly, 57% have chronic insomnia and only 12% report no sleep complaints.
- Only 5% of insomniacs make a visit to the doctor specifically for a sleep problem.
- Only 26% discuss sleep problems during a visit for another purpose.
- 69% never discuss it at all.

The prevalence of insomnia is very high in the following medical conditions which are well known to cause insomnia[20]

- Pain: angina, arthritis, hip replacement, cancer
- Neurologic and psychiatric disorders: restless leg syndrome, neurocognitive impairment (NCI), AD, Parkinson's disease, delirium, anxiety, depression, drug abuse and withdrawal
- Organ system failures: diabetes, heart disease, heart failure, angina, asthma, chronic obstructive lung disease, gastric reflux, prostate enlargement, incontinence

Impairments associated with insomnia[21]

- Impaired cognitive functioning
- Deterioration in quality of life
- Increased body pain and poor general health
- Higher risk of future psychiatric disorders
- Increased risk of absenteeism and poor job performance
- Increased risk of accidents
- Increased health care costs

Watch what you ingest!

There are many over the counter and prescribed medications that cause insomnia as a side effect. Some substances, like alcohol, coffee, tea and chocolate are even considered foods. Most juices and sodas have caffeine added to them. Even some bottled waters and some beers may have added caffeine. Read the label! While it may feel great to have a cup of coffee in the morning, added caffeine during the day will have unwelcome effects. Caffeine has a long half-life and stays in the system for many hours. Some coffees contain too much caffeine – guess which one! Remember, caffeine is an addicting substance and caffeinism is a disease!

No magic pill

Sleep is a regular activity and like all regular activities, it is also a habit. The best way to achieve regular refreshing sleep is through the avoidance of counterproductive practices and the establishment of good *sleep hygiene* – which should become a strong habit.

Proven, non-pharmacological ways to establish sleep hygiene and to improve sleep disorders

- Get up at the same time every morning
- Go to bed only when you are sleepy
- Give yourself about 20 minutes to fall asleep
- Allow time to wind down (2-3 hrs) – avoid stimulating activity (TV, work, excting book, etc.)
- Keep regular routines and rituals before bedtime (e.g. tidy room, brush teeth, shower, bath, etc.)
- Exercise moderately up to afternoon but not in the evening
- Have only a light protein snack or warm milk at bedtime
- A warm bath then cooling down induces sleep
- Keep the bedroom dark, quiet and and cool (but not enough to feel cold)
- Avoid caffeine, alcohol and tobacco in the evening
- Don't oversleep
- Save your bed for sleep, sex, sickness (SSS)
- Avoid regular use of sleeping pills

If establishing good habits that facilitate the onset and preservation of sleep does not work, there are a number of nonpharmacologic therapies that have been found to help insomnia. These include cognitive behavioral therapy (CBT), relaxation therapy, stimulus control and paradoxical intention.[22]

There is no such thing as an ideal sleeping pill. Most *hypnotics*, as sleeping pills are known, tend to work only for a limited time, sometimes only for a few weeks, and many have serious side effects. Some will depress the nervous system causing fatigue and cognitive impairment. Some will impair balance and cause falling. Some are addicting, so that you will need to take more and more to get the same effect and withdrawal will lead to high anxiety and rebound insomnia. Some will affect blood pressure. However, some people do tolerate and need sleeping medications on the long run. Pharmacological therapy should be the last resort but if necessary, it should be used because chronic sleep deprivation may be a lot worse than expertly prescribed and monitored hypnotic medication.

THE LEAST YOU SHOULD KNOW

- Sleep deprivation is rampant but it does not have to include you.
- Sleep is a complex brain activity.
- Sleep is vital to optimal cognitive functioning.
- Sleep is vital to normal hormonal functioning.
- Sleep is vital to immune system functioning.
- Chronic sleep disorders and sleep debt will make you sick.
- Sleepy people are dangerous to themselves and others.
- Sleep disorders must be diagnosed and treated – safe treatments are available.

START NOW!

- Use the tools supplied on the next pages to improve your sleep efficiency.
- Establish the habit of good sleep hygiene.
- Educate others about the importance of refreshing, drug-free sleep.
- If you have a sleep disorder, be assertive in getting a diagnosis.
- Get treatment, not just diagnosis of sleep disorder.
- Use non-pharmacological treatment methods first.
- If possible, avoid long term use of hypnotics.

www.HealthyBrain.org

Sleepiness Screening Test[23]

How likely are you to doze off or fall asleep – not just feel tired – in the following situations?

This refers to your usual way of life in recent times. Even if you have not done some of these things recently, try to work out how they would affect you.

Use the following scale: 0 = never; 1 = slight chance; 2 = moderate chance; 3 = high chance

Situation	**Chance of dozing**			
Sitting and reading	0	1	2	3
Watching TV	0	1	2	3
Sitting still in a public place (e.g. theatre, meeting)	0	1	2	3
As a passenger in a car for an hour, without a break	0	1	2	3
Lying down to rest in the afternoon	0	1	2	3
Sitting and talking to someone	0	1	2	3
Sitting quietly after a lunch without alcohol	0	1	2	3
In a car while stopped for a few minutes in traffic	0	1	2	3

Total score: _____

If your score is seven or higher, you have sleep debt. This means that you are not getting the sleep you need. It may be because of sleep apnea or some other factor. Many conditions can cause excessive daytime sleepiness. Please see your physician.

To calculate your Sleep Efficiency measure your time asleep and divide it by the time spent in bed. Multiply that by 100 to get your Sleep Efficiency (SE) as a percentage.

SE = Time Asleep/Time in Bed x 100

SE should be 80-90 %.

If it's too low you are not sleeping but spending much time in bed. Does that mean you are depressed and do not wish to get up? Are you just lazy? Are you unable to sleep and tired in the morning despite trying to sleep?

If it is too high you are so sleepy because you are not getting enough. You are not 'falling asleep', you are 'crashing' and you may be sleepy in the morning. You may be exhausted running a sleep debt.

www.HealthyBrain.org

CHAPTER 12

STRESS

I burn my candle at both ends;
It may not last the night.
But oh! My friends, and ah! My foes,
It makes a lovely light!

– Unknown

What is stress? Non-medical and non-scientific notions about stress are reflected in the common Oxford dictionary definition, which makes mention of "hardship", "affliction", "force upon another" and "strain on mental powers". Those rudimentary ideas are rooted in what the ancients believed. Hippocrates (500 BCE) related it to *pathos* and *ponos*, in Greek meaning 'suffering' and 'toil'. Hippocrates intuitively concluded that pathos and ponos caused "humoral imbalances". It was not until around 1850 that Claude Bernard, perhaps the greatest pioneer researcher in physiology, discovered what he called the *internal milieu*, our inner fluid world of salts, nutrients and other chemicals that, as he emphasized, must stay constant otherwise we get sick and death soon follows. Claude Bernard conceptualized that stress caused a disturbance in the *internal milieu*.[1] The word 'stress' or 'stressor' did not come into use until much later. You will learn about this when we discuss the discoveries of the "Father of Stress Medicine", Dr. Hans Selye.

Everyone knows that a little stress is a good thing. This is true – a bit of stress seems necessary even to stay alert – to perform at all. Too much is devastating. We have some irrational beliefs about stress, such as: 'you can't avoid stress', or 'what doesn't kill you makes you stronger'. Bear with me a bit longer and I will explain how the stress circuits work in your brain and body and what happens when stress triggers push the system out of control. As a primer, think back to the 'inverted U' curve that we discussed when we addressed Mental Activity in Chapter 10. Things that cause stress and the effects of stress follow a similar pattern. A 'little bit' of most things in daily life have little effect, the right amount is the best (optimal), and 'too much of something' becomes stressful. (e.g. noise,

161

heat, etc). That was Dr. Hans Selye's definition of stress – that it was "too much of something". Also the 'inverted U' relationship holds for the effects of stress. Too little and we lose concentration, the optimal amount, at the top of the 'inverted U' gives best performance, too much stress and we become distressed and unable to function. The stress response is triggered.

There are innumerable stress triggers. Some of them are physical, such as noise, heat, cold. Some are physiological, such as malnutrition, or sleep deprivation, or too much sensory stimulation. Others are environmental such as crowding or isolation. Still others are psychological such as loneliness, loss or fear. Some stress triggers, contrary to common belief are actually things often thought of as pleasant. Winning the lottery, for example may seem pleasant but people who win report experiencing significant stress in their lives. Your wedding day may be the happiest of your life, but it can also contain some extremely potent stress triggers. How do you know if something is a stress trigger? The answer is simple. An event is a stress trigger, or stressor, if its presence causes the body's stress system to activate. It is as simple as that. If it doesn't cause the body to go into stress mode, it is not, by definition, a stressor. If you are now thinking "aha", that is why some people seem really stressed out in a situation while others seem calm and mellow – some have activated stress systems and others have not – you are absolutely right. Stress is a relative concept; one man's stress is another man's mild stimulation or even pleasure.

Another phenomenon that we must introduce here is that of *adaptation* to stress. Studies show amazing adaptation both in animals and humans. For example, the first time parachutists jump their stress hormone levels are very high by the time they reach the ground. This response diminishes over subsequent jumps and eventually jumping out of an airplane does not register a big stress reaction.[2]

Discussions on stress may seem puzzling because we have created too many definitions or concepts of stress, primarily by confusing cause and effect. If you accept the position that a *stressor (stress trigger)* is the stimulus and the *stress response* is what we experience in the presence of a stressor stimulus, or trigger, the whole topic will become clearer.

Body-brain reactions

Now that we have determined the difference between a stressor/trigger and a stress response, I will turn to the more important business of describing how the body and the brain react to a genuine stressor. Notice that I have not said how the mind reacts. That would be getting way ahead of the game. My first task is to locate the stress response firmly in the body and brain. I will discuss cognitive effects of stress later.

It turns out that one of the most important medical discoveries of the mid 20th century was the *stress response*. Yes, it took until the mid 20th century for us to understand how the body responds to the presence of a stressful substance or event. Until Dr. Hans Selye took on the task, no one even thought of stress as a unified concept. Selye turned that situation around when his studies of animals exposed to physical stressors, noise, mild electrical shock, etc., showed characteristic patterns of arousal and, with continued exposure to those stressors, characteristic physical problems including heart disease and digestive disorders.

Although Selye's work excited the medical community and general population, its novelty was such that it required certain changes in the culture. For example, biographers of Selye noted that when he was asked to present a paper in France for the first time, there was no equivalent word in French for *stress*. That created an obvious problem which was quickly solved when his French hosts coined a new word *le stress*, which allowed Selye to introduce his work to France. Not to be outdone, the Germans, who have also invited Selye to speak and who, like the French, had no word in their language corresponding to *stress*, immediately coined a new one: *der stress*. We can think of no greater acknowledgement than that of changing a language to accommodate a concept!

Hans Selye – the Father of Stress Medicine

Hans Selye was born in Vienna in 1907. While still in Medical School he became interested in how the presence of disease affects people's general physical function and well being. His most important finding was that patients who suffered from a variety of different medical problems showed many of the same symptoms. Even more important was his explanation, which stated that the similarity in the symptoms was a direct result of the physiological response to a stress factor – the presence of illness. This was a radical concept for the time and was to have far-reaching consequences. Selye called this phenomenon the *stress syndrome*, and later the *General Adaptation Syndrome* (GAS).[3] Selye spent much of his adult life in Montreal, Canada where he founded the Canadian Stress Institute. His theories and research formed the basis of modern stress medicine and have had a profound impact on both medical practice and popular culture.

The General Adaptation Syndrome (GAS)

Selye found that the body's response to the presence of a stressor depended on the severity of the stressful event and the length of exposure.[4] If the exposure was brief, the body would quickly return to normal function. More importantly, he found that if the exposure was prolonged, as when people were sick for extended periods, the body went through three distinct physiological stages.

Stage One: The Alarm Response

The *Alarm Response* is the classic *fight or flight* response. When the body is exposed to a stressor it immediately mobilizes to deal with it. If it is an external stimulus, such as a bear sticking his head in your door, you either run away or try to chase the bear away. During this time, brain chemicals such as adrenaline, noradrenaline and dopamine peak and the adrenal glands pour out more adrenaline into the system to prepare the body for 'fight or flight'. Either way it is over in a brief period of time, then you go back to your porridge, or chair, or even your bed.

Stage Two: The Adaptation Phase

If the stressor continues to be present, the body moves into a new mode of function by changing its metabolism to allow for prolonged effort, whether it be for fighting off more bears, or running farther away, or in the case of illness, fighting the disease or infection. This requires substantial energy and consequently cannot last indefinitely. During *Adaptation Phase* you may not need to be extremely vigilant for new events but much energy may be needed to cope. Injury and exhaustion may be ongoing. Brain chemicals change. Dopamine drops but adrenaline remains high and cortisol, the real stress hormone, elevates.[5] This starts the damage – cortisol is the hormone of death!

Stage Three: Exhaustion

If the stressor continues, eventually the body runs out of resources and collapses. That occurs when the cost of keeping the body at a high damage rate exceeds the body's capacity for renewal. As this occurs, all physiological systems are affected and slowly but inexorably collapse. Very high cortisol levels initiate emergency physiological changes to prepare the body for severe injury and impending death. The changes include protein breakdown, immune system shut down, increased blood clotting, and lower pain perception.

The General Adaptation syndrome (GAS) forms the basis for our understanding of how exposure to stress leads to predictable consequences. Since Selye's original work, many talented researchers have made great progress in mapping out the stress system in the body and the stress circuitry in the brain. I will explain these important brain chemicals and hormones later. Just keep in mind the names dopamine, adrenaline, noradrenaline and cortisol for now. Those neurotransmitters are responsible for the acute effects that are more easily recognized than chronic changes. These can be grouped as physical, emotional and cognitive effects, which can cause very unpleasant symptoms.

Bodily changes associated with symptoms of the acute stress response – the Alarm Response

Physical	Emotional	Cognitive
increased heart rate	moodiness	forgetfulness
sweaty palms	nervousnesss	poor concentration
headaches	irritability	poor judgement
cold hands	anxiety	disorganization
indigestion	distress	fuzzy perception
nausea	unhappy	confusion
shortness of breath	no sense of humor	lack of interest
holding breath	labile (jumpy) mood	math errors
disturbed sleep	sadness	overfocus on one thing
pain	defensiveness	racing thoughts
fatigue	anger	negative self-talk

The above is a partial list and the symptoms are not in any particular order, as they may appear singly or in various combinations. The point is to show you that there are many objective and subjective events that can be stress symptoms. Some of the are easily measurable (e.g. blood pressure), some are not. Stress symptoms can be extremely unpleasant and may even precipitate panic, which is an aimless swing into action to reduce tension. Below is a summary of the neurotransmitter profiles for various stages of the stress response.[5.1]

	Dopamine	Noradrenaline	Cortisol
Alerting response, anxiety	extremely high	moderate	high
Coping, adaptation	moderate	extremely high	very high
Depression, exhaustion, illness	moderate	….very high	extremely high

The Stress Circuit

If you want to understand stress and the stress response, you need to know about the stress circuit. This is a cascade of biochemical events that start and perpetuate and eventually stop the stress response. The term stress circuit is a short name for the limbic system, hypothalamus, pituitary gland and adrenal gland (or LHPA axis, to be even shorter). The LHPA, often referred to simply as the HPA axis, is a term we use to describe a complex feedback system that links the brain, and other organs in the body, in a loop, where stress triggers (internal or external) cause the release of neurotransmitters called catacholamines and the hormone cortisol. Consider the following chain of events:

Information → brain → limbic system → hypothalamus → releasing factors → anterior pituitary gland → adrenicorticotrophic hormone (ACTH) through blood → adrenal cortex → cortisol (into blood) → organs

The catecholamines, the neurotransmitters active in the stress reponse, include the following:

- Dopamine – increases alertness and vigilance. Focuses attention within narrow limits.
- Adrenaline (epinephrine) – increases the availability of glucose in the blood which will be needed by the brain and muscles. It also activates the brain.
- Noradrenaline (norepinephrine) – increases heart rate and blood pressure to supply increased flow to muscles. Increases muscle tone.

All of the above help to trigger the increase in the stress hormone cortisol in a progressive manner. When you experience a stress trigger, as when the bear stuck his head in the window of your house, the limbic (emotional) system reacts strongly and the hypothalamus (a control structure in the brain) releases a hormone, dopamine, which hyper-alerts the brain and focuses it one one thing – the bear. At this initial point, adrenalin and cortisol are at lower levels. At the same time a big signal is released to the pituitary gland (located at the base of the brain) which in turn releases a second hormone into the blood stream. That hormone serves to activate the adrenal glands, two structures that are located at the top of your kidneys. The adrenal gland response takes a few seconds longer but then the action really begins. When the adrenal glands are stimulated by the pituitary gland, they immediately begin to secrete and release several hormones and hormone compounds. The most significant of these are more dopamine, adrenaline and noradrenaline. The stress hormone cortisol is next to escalate.[6] Let's look at what happens in even more detail.

The response of the second stage, after the initial Alarm Stage, is the Adaptation Response, which is really a coping response. Dopamine drops, but adrenalin and noradrenaline stay high. Cortisol elevates preparing the body for a more protracted engagement. Adrenaline and noradrenaline are the neurotransmitter hormones known for giving the body a quick boost of energy and strength. They do this by increasing your heart rate and blood pressure and causing blood to be diverted away from the stomach and internal organs, to your muscles. It also makes glucose available from the liver. Those events are the basis for the increased strength, stamina and improved reaction time that people with elevated levels of fight-flight hormones experience. Dopamine is the neurotransmitter that starts the whole cascade of events because dopamine regulates attention and mental focus. The end result of the cascade is an outpouring of cortisol, the real heavy duty stress hormone. During protracted exposure to the above stress hormones, the body and brain must adapt and cope. The changes over weeks, months or years during the Adaptation Response result in chronic diseases (e.g. hypertension, diabetes, etc.). That is how the body copes until exhaustion and organ failure.

Neuroendocrinologists think of cortisol as a stress hormone, but we also take it very seriously. One of the interesting effects of cortisol is to release even more glucose from its stored state into the blood stream. As the stress (say bear in the house and struggle continues), past the coping phase, cortisol prepares for the 'end game'. That is why I referred to it as the death hormone earlier in this chapter. It sets the body for exhaustion, depression and illness. It also causes salt and fluid retention and a thickening of the blood making it prone to clotting. It shuts down a number of reactions in the body especially those responsible for growth, reproduction and the immune system. If you think back to the discussion on the brain as an organ, in Chapter 4, you will recall that glucose is the basic fuel both for the body and the brain. So the first action of cortisol (along with adrenaline) of the stress response is to flood the bloodstream with fuel to power both physical and brain metabolism – the other effects are to prepare for more deadly outcomes, such as physical injury. Cortisol prepares the victim to die – slowly. That is why cortisol has been nicknamed the 'Frankenstein hormone'. It means well and seems to do no harm and it is even beneficial in small, quiescent, appropriately timed, amounts, but when it gets strong and goes haywire – watch out!

Cortisol also has a very important secondary role. When a stress trigger is removed, and the stress circuit needs to be turned off, cortisol has an inhibitor effect on the hippocampus and related brain regions (hypothalamus), effectively turning it off and ending the stress response. This negative feedback loop is very important. When it is impaired, cortisol does not return promptly to lower normal levels and bodily, including brain, damages can occur.

The hormones produced by the HPA axis in response to stress triggers also have an impact on both brain and the endocrine system. In the brain, stress hormones influence the limbic system which comprises the amygdala, hippocampus and cortex of the brain. The limbic system, also known as the emotional brain, is intimately involved in motivation, emotions and the perceptions of pleasure or danger. That explains the strong feelings and intense responses people experience in the presence of powerful stress triggers.

The hippocampus is intimately involved in the processing of information and formation of memory. We know that when people experience intense stress triggers they may form very strong and persistent memories of those events; memories that sometimes persist and can be intrusive (e.g. PTSD). This is likely a consequence of the over-activation of the hippocampus while in a highly aroused physical and emotional state. For reasons we do not understand, cortisol has a toxic effect on brain cells particularly hippocampal cells.[7] In the hippocampi, the part that regulates new learning and retrieval of information from the rest of the brain, cortisol wreaks havoc. The hippocampi (we have two of them, one on each side just medial to the temporal lobes) actually start to shrink when exposed to high cortisol levels for prolonged periods. No wonder chronically stressed people have trouble with attention, concentration, learning and remembering.

We also know that the stress response produces changes in appetite, and other metabolic processes. Basically, the stress response switches the body and brain from normal to emergency function, so sensitivity to hunger, temperature, pain, pleasure and attention to detail are diminished, presumably because these functions divert attention and energy from short-term survival needs.

Finally, it is well established that the brain chemicals, hormones and hormone substances, secreted and released during a stress response change the functioning of the immune system, the reproductive system and metabolism. The above changes follow a common theme – they prepare the body and brain for an immediate and brief potentially catastrophic event at the expense of normal, healthy metabolic processes.

Stress and brain health

Now that you know a considerable amount about the nature of stressors and have grasped the concept of the stress response and understand the basic wiring of the stress circuit (HPA axis), we can turn to the primary topic, stress and brain health.

As mentioned in the previous paragraphs, stress has both short and long term effects on the brain. If we go back to the three stages of the General Adaptation Syndrome, we can make some comments about how the brain responds to stress.

Short-term brain changes

In Stage One – the Alarm Response – the brain, like the body, goes into emergency mode. The increased glucose in the blood stream gives a metabolic boost, the adrenaline in the system increases arousal and the cortisol changes your patterns of emotions and cognition. Simply speaking, the brain becomes hyper: hyperfocused, hyperalert and hyperresponsive. One of the first parts of the brain to be affected by a stress response is the hippocampus. Think of it this way: When the bear sticks his head in the door, you don't need to be using the parts of the brain that sort bears by genus and species. The fact that it is an 'ursus horribilis' (the common grizzly) seems hardly relevant and it is not surprising that those parts of the brain involved in such intellectual pursuits turn down the volume or even become completely dysfunctional. However, you will be wanting to pay very close attention to the bears movements and facial expressions as this may well be your only clue to whether you fight or flee. Of course, this all happens in a flash – before you can think! So thinking goes down, attention goes up. As mentioned, this is largely the action of the alerting brain chemical called dopamine. The available research, and the accounts of many people who have experienced severe stress, indicate that something profound happens to memory function as a direct result of the presence of high levels of stress hormones. Normal memory function becomes less efficient and in many cases people will have fragmented or partial recall of the stress producing events. On the other hand, they will also often have very vivid and detailed memories of certain aspects of the events. It is as if memory processing is disrupted for all but certain highly relevant aspects of the situation.

The limbic (emotional) brain system also goes into atypical function as well. The hippocampal system, linking the limbic system to the memory system, shuts down and the two motivational-anxiety-fear centres, the amygdala, fire up. Thus, emotional responses and motivation are greatly increased. That may not be surprising. A good boost of motivation and a healthy shot of fear or anger will certainly fuel the fight or flight response. During that hectic time you would have a great deal of difficulty remembering facts, even simple important phone numbers. There is a reason why emergency numbers are simple and short.

These types of short-term brain responses are clearly adaptive, as it makes good survival sense to develop strong memories of dangerous things and to discard irrelevant information. Our marvelous brain can launch the process in a fraction of a second. If you run into similar dangers, the brain will set off an alarm – the highlights you will surely remember. However, this very adaptive process can go somewhat out of control and lead to the development of post-traumatic stress problems that are characterized by strong, unpleasant, persistent and intrusive memories. Such deeply embedded memory fragments or impressions of traumatic events are sometimes extremely troublesome because they can be inadvertently triggered as vivid emotional responses and can be intrusive and disruptive, or even severely distressing, in their own right – that is the *Post Traumatic Stress Disorder* (PTSD). The vivid memories and feelings are described as 'flashbacks' by those suffering from PTSD. PTSD is a serious condition that requires diagnosis and treatment.

Long-term brain changes

As mentioned, the 'adaptation' and 'exhaustion' stages of the GAS are short lived and the response shifts towards less dopamine and more adrenaline, noradrenaline and cortisol. The chronic activity of those neurotransmitters cause brain changes and organ damage – systems of the body begin to suffer. The brain is not immune from this process and modern neuroscience research has demonstrated that deleterious changes occur in both brain structure and brain function when animals and people are exposed to high levels of stress for long periods of time.[8] The hippocampus is the first brain structure to be affected by chronic stress. Very good studies by the American Army of soldiers exposed to combat clearly show that the hippocampus begins to shrink when soldiers experience combat and that the longer they are exposed, the greater the loss of tissue.[9] The neuroscientists who have examined this phenomenon believe that the release of stress hormones and cortisol floods the brain and has a toxic effects on hippocampal cells. If the cortisol levels remain high, the cells begin to degrade and die. That lowers the actual volume and efficiency of the hippocampi with a resultant loss in the ability to process information and memories. Fortunately, when the stress is removed the hippocampus begins to regenerate.[10]

The second structures to be affected by chronic stress are the amygdala, which, as noted before are heavily involved in emotional states, fear and motivation. In the case of the amygdala, researchers are finding that the presence of high levels of stress hormones leads to increased size and a more complex cellular structure in the amygdala.[11] That suggests that one effect of prolonged exposure to severe stress is increased fear, rage and emotionality – exactly the sort of things that are observed in people who are identified as suffering from PTSD.

The brain is also indirectly at risk because of the increased likelihood of heart attack or stroke that is is associated with prolonged exposure to chronic stress. That too is a side effect of chronic stimulation with adrenaline and cortisol. We tend to think that there are different kinds of stress. We speak of executive stress, doctor stress, caregiver stress, financial stress and others but the final common pathway is the same.

In summary, the effects of chronic stress include:

- Increased abdominal fat
- Increased blood viscosity

- Deccreased blood flow to the gut and other visera
- Change in appetite (increase or decrease)
- Weight gain or loss
- Sleep disorder
- Comprehension or memory problems
- Depression
- Hypertrophy of the amygdala
- Atrophy of the hippocampus

Animal experiments

The fact that stress can cause physiological and even structural bodily changes alerts us to the possibility that there are serious implications for brain development. Experiments and observations on animals strongly suggest that we should be very concerned about early parenting.

Early emotional stress, in the form of parental deprivation, disrupts synaptic growth and brain development. The neuroanatomy of parental deprivation has been well studied. Experimental findings show that when newborns – in this experiment degus pups (chinchilla-like rodents), were separated from their parents, they subsequently showed neuroanatomical changes.[12] These results support a neurobiologic basis for the behavioral syndromes found in rodents, nonhuman primates and humans after early maternal separation. The atrophy identified in limbic regions are consistent with findings of separation trauma associated anatomic and neuroimaging differences in the limbic systems of both humans and animals.[13]

Another study showed that stress levels were higher amongst the timid. In this interesting study, investigators studied constitutionally *timid* compared to *sociable* baby rats. Constitutionally timid animals responded to novel situations with significantly greater increase in behavioural inhibition and stress hormone levels than sociable animals. Basal stress hormone levels and rate of recovery were similar but elevated levels were prolonged in the timid group because of their higher peak levels. Timid animals died significantly earlier than their sociable counterparts.[14]

In another study, researchers divided a set of rats into two groups. One was allowed to live communally (the no stress condition). The rats in the other group lived alone (the stress condition) during adolescence. Ethologists verify that for rats, living alone is highly stressful. They then examined the brains of the group members and found that the rats who had lived in the stressful environment showed a significant decrease in a specific protein found in the hippocampus. The protein was a promoter of the formation of new synapses (synaptophysin), which would normally increase during that period of life.[15]

In short, animal studies confirm the observation that early exposure to stressors leads to changes in adult behaviours and even in truncated longevity. Results of animal experiments are certainly consistent with observations on human childhood and adolescent development. Although the researchers are cautious in their interpretation, the results suggest that high stress environments are particularly bad for brain development.

Human studies

The effects of severe stress are difficult to study in humans. Who after all would undergo such experiences to show that the hippocampus gets degraded after stress? Adolescent stress is an issue under scientific scrutiny. Can exposure to stress in adolescence lead to brain changes in adulthood? Studies by Susan Andersen, PhD, of McLean Hospital in Massachusetts have shown that stressful events experienced during adolescence can lead to enduring changes in brain processes expressed in adulthood.[16] In another salient study, it was shown that the hippocampi atrophy during clinical depression and they continues to do so as long as the depressive illness continues.[17] We have quite a bit of data on stress effects on the population. Did you know?

- Awakening is stressful and is associated with increased risk of cardiovascular events, whether awakening in the morning or from a nap.[18]
- Work stress has been shown to be a risk factor for a major depressive episode.[19]
- The British Department of Health and Meteorological Office Collaborative Study found that the risk of death increased after a temperature drop as follows: coronary thrombosis = 2 days, stroke = 5 days, chest infection = 12 days.[20]

You may be interested in some other stress facts from Canada and the USA:

Stress facts – Canada

The Canadian Consumer Survey on Health Care[21]

- Distressed by work: women = 53%; men = 41% (up 33% over a two year period)
- Public sector = 53%; Private sector = 42%
- Work overload = 43%
- Financial worries = 35%

- Balancing work & family = 32%
- Caregivers of people with major neurocognitive impairment (dementia): depression = 40%
- Caregivers of those without major cognitive impairment (non-dementia): depression = 20%

Canadian Employers Report[22]

- 83% cite stress as key employee health problem
- Wellness program pays 6:1
- Only 17.5% of employers offered a wellness program

High Risk Groups:

- The poor and those in debt
- Physicians
- Single moms
- Post-menopausal females
- Adolescents
- Caregivers – especially those caring for victims of brain disease
- Elderly in long-term care (LTC – i.e. nursing home) residence

Stress facts – USA

- US Department of Health Services reported that 75% of the general population experiences at least some stress every two weeks. Half of those people experience moderate or high levels of stress during the same two-week period.[23]
- Millions of Americans suffer from unhealthy levels of stress at work. An early study reported several years ago estimated the number to be 11 million. Given events since that time, this number has certainly more than tripled. Studies in Sweden, Canada, and other Westernized countries show similar trends.[24]
- Worker's compensation claims for mental stress in California rose by 700% in the 1980s, whereas all other causes remained stable or declined.[25]
- *Prevention of Work-Related Psychological Disorders* was a National Strategy proposed over fifteen years ago by the National Institute for Occupational Safety and Health (NIOSH).[26] The US Public Health Service made reducing stress by the year 2000 one of its major health promotion goals. We are way past the deadline.

- Stress contributes to heart disease, high blood pressure, strokes and other illnesses in many individuals. Stress also affects the immune system, which protects us from many serious diseases. Stress also contributes to the development of alcoholism, obesity, suicide, drug addiction, cigarette addiction and other harmful behaviors.[27]
- Tranquilizers, antidepressants, and anti-anxiety medications account for one-fourth of all prescriptions written in the USA each year.[28]

In 1997, Robert Sapolsky said, "I can imagine few settings that better reveal the nature of psychological stress than a nursing home". Have things changed?

Doctor stress

An Irish intern worked for 12 hours straight before he collapsed having worked 100 hours per week prior to collapse. The diagnosis was "exhaustion on job". After the collapse he was hospitalized. When he returned he was asked to do an additional 48 hrs on call, lost 2 days pay and he was billed by the hospital! A subsequent Irish Medical Organization study showed that the average doctor worked 77 hours per week and many worked 120 hours per week.[28]

Death is often stressful

Studies have shown that cortisol levels are twenty-fold higher in the cerebrospinal fluid (CSF) from deceased subjects than in CSF from living subjects. That suggests the experience of impending death, regardless of end of life interventions that seek to mitigate its trauma and agony, provokes a stress reaction with physiological and possibly psychological contributories. This includes patients with severe neurocognitive impairment (referred to as dementia in the study) and also those receiving morphine. Neither the severe cognitive impairment nor morphine were found to be protective against the elevation of stress hormone. CSF cortisol levels in all subjects were fifteen to twenty times higher than those commonly seen in living subjects. Morphine dosage was unrelated to cortisol levels. Differences in time and season of death, post mortem duration, CSF storage, time before testing and age at death were unrelated to cortisol levels. It is interesting to note that elevated cortisol levels have been observed in living Alzheimer's patients but these levels are much smaller, being up to about 2 times higher.[29]

Given the above rude awakening, we should be very concerned about stress management in our daily lives and in the care of the dying.

Identifying sources of stress in your life

One of the keys to keeping stress from controlling your life is to identify the stress factors that start the stress response. Remember, stress is the source of your problems, not your state of mind. This means that a major part of managing stress is figuring out what your stressors are. If you can control them, you are way ahead of the game. In the world of engineering, this is the key to building structures that are stable. In life, it is a key to staying healthy. A stress trigger is any event that activates the HPA axis and turns on the stress circuit. I cannot strongly enough emphasize the word 'any'. In our modern self-help world we have put too much emphasis on the psychological stress triggers. If you look at any of the frequently visited internet stress sites, you will immediately see that they tend to focus on emotional and social events that complicate peoples lives. It is not that they are wrong. Losing your job, experiencing a death in the family etc. are clearly stress triggers. However, they are only one type of stress trigger. They also tend to emphasize the negative events of life. Again, while this is not wrong, it hides the fact that 'positive' events can also be stressful. A promotion, retirement or winning a large sum of money are thought of as positive events but such events are known to be real stressors – they activate the stress circuit. To make our point, events that are not psychological but physical also register as stressors and often severe ones at that. Even eating a large meal elevates cortisol levels. So does chronic alcohol consumption, surgery, physical injury, pain and the use of a multitude of drugs – to name just a few common stressors.

Sample stress triggers

Physical	Social	Psychological
Heat	Crowding	Loss
Cold	Isolation	Fear
Pain	Aggression	Change or novelty
Noise	Chaos	Uncertainty

Illness is stressful. That is not a surprise. (One more reason to stay healthy.) What is surprising, however, is that, judging from the huge elevations in cortisol levels, the end of life appears to be associated with severe stress. Does this have to be so? Why does it happen? We don't know. Many always assumed that severe neurocognitive impairment (dementia) was somehow protective against the stressors we experience. The rationale for this was that if one cannot understand or remember anything one must be in a kind of ignorant bliss – nature's way of being kind. It may be baloney!

No discourse on stress would be complete without the Holmes and Rahe Stress Scale. In a now famous American study from 1967, Dr. Thomas H. Holmes and Dr. Richard H. Rahe developed a do-it-yourself stress test called the *Social Readjustment Rating Scale*. This is a very useful little tool to evaluate the presence of a potentially stressful life and social events. Remember, there are many stressors or stress triggers that are not social. However, as this list shows, many aspects of life are potentially stressful and Holmes & Rahe approach is a simple way to identify some of them.[30]

I like it because it emphasizes that some events are more stressful than others. It also highlights the point, and I fully agree, that even things that seem positive can be stressful and you can never be sure which way things will go. If you get that promotion will it be good for you? Only time will tell.

Life event Life Change Units

Life event	Life Change Units
Death of a spouse	100
Divorce	73
Marital separation	65
Imprisonment	63
Death of a close family member	63
Personal injury or illness	53
Marriage	50
Dismissal from work	47
Marital reconciliation	45
Retirement	45
Change in health of family member	44
Pregnancy	40
Sexual difficulties	39
Gain a new family member	39
Business readjustment	39
Change in financial state	38
Change in frequency of arguments	35
Major mortgage	32
Foreclosure of mortgage or loan	30
Change in responsibilities at work	29
Child leaving home	29
Trouble with in-laws	29
Outstanding personal achievement	28
Spouse starts or stops work	26
Begin or end school	26
Change in living conditions	25

Revision of personal habits 24
Trouble with boss... 23
Change in working hours or conditions 20
Change in residence... 20
Change in schools... 20
Change in recreation.. 19
Change in church activities 19
Change in social activities.. 18
Minor mortgage or loan.. 17
Change in sleeping habits.. 16
Change in number of family reunions.......................... 15
Change in eating habits 15
Vacation.. 13
Christmas.. 12
Minor violation of law.. 11

To do the above test, check each item that you have experienced in the past 12 months and add them up. Research by Holmes & Rahe suggests that a score of less than 150 is a *minor stress*. Those who score 150-300 are experiencing *moderate stress*, and a score over 300 represents *major stress*. It is estimated that 33% of those with a score below 150 will experience an illness or accident, while those with a score between 150 and 300 have a 50% chance, and those with a score over 300 have a 90% chance of a significant illness or accident within two years.

Stress management

Stress management that focuses on avoiding stress is misguided and cannot succeed. Stress management should more appropriately be called *avoiding being stressed*. There are three distinct parts to stress management:

Part one: Learning to notice the signs of onset for a stress response.

Part two: Identifying and managing the less obvious stress triggers and stressors.

Part three: Understanding and managing your personal response and doing what works for you.

Most books and articles on stress management focus on Part Three, managing your response. I disagree. Trying to manage stress without dealing with the immediate onset is like ignoring a fire. Ignoring stressors is like trying to manage burns without dealing with the fire. You absolutely need to manage both.

Part one – learning to notice the signs of onset for a stress response

Recognizing the onset of a stress response is easy. Acute stress, when above a certain threshold is very uncomfortable. Recall the table of physical, emotional and cognitive symptoms. Who can live with those? When we are stressed so severely, the solution seems obvious. Escape, complain, put a stop to the process, scream! If you can't, you may be toast. The obstacle is that some stressors are not so severe as to produce all those nasty, easily recognized effects. The real problem is chronic stress.

Part two – identifying and managing the less obvious stress triggers and stressors

When the acute stress response is mild and if it is ignored, the more chronic form is ushered in. You can get used to excessive stress and when this occurs the symptoms become more difficult to identify. Just because you don't feel the activation of acute stress does not mean that the stress response is altogether absent – you just do not notice it any more. We cannot tell what our cortisol levels are any more than we can guess our blood pressure or sugar levels. How do you know whether you are or are not stressed? For this reason, there are tools and guidelines for identifying chronic stress. Unfortunately it is not as easy and accurate as measuring blood pressure. Blood cortisol measurements are not reliable. Salivary cortisol levels would be more accurate but they would have to be done serially to get a summation. We have no accurate clinical tests for chronic stress.

Test your vulnerability to stress

In modern society, most of us can't avoid stress but we can learn to behave in ways that lessen its effects; we can modify our reactions to stress. Researchers have identified a number of factors that affect one's vulnerability to stress – among them are eating and sleeping habits, caffeine and alcohol intake, and how we express our emotions. The following questionnaire is designed to help you discover your vulnerability quotient and to pinpoint trouble spots.[31]

Rate each item from 1 (always) to 5 (never), according to how much of the time the statement is true for you. Be sure to mark each item, even if it does not apply to you – for example, if you don't smoke, circle 1 next to item six.

1.   I eat at least one hot, balanced meal a day.   1 2 3 4 5
2.   I get seven to eight hours of sleep at least four   1 2 3 4 5
     nights a week.

3. I give and receive affection regularly. 1 2 3 4 5

4. I have at least one relative within 50 miles on whom I can rely. 1 2 3 4 5

5. I exercise to the point of perspiration at least twice a week. 1 2 3 4 5

6. I limit myself to less than half a pack of cigarettes a day. 1 2 3 4 5

7. I take fewer than five alcohol drinks a week. 1 2 3 4 5

8. I am the appropriate weight for my height. 1 2 3 4 5

9. I have an income adequate to meet basic expenses. 1 2 3 4 5

10. I get strength from my religious beliefs. 1 2 3 4 5

11. I regularly attend club or social activities. 1 2 3 4 5

12. I have a network of friends and acquaintances. 1 2 3 4 5

13. I have one or more friends to confide in about personal matters. 1 2 3 4 5

14. I am in good health (including eye-sight, hearing, teeth). 1 2 3 4 5

15. I am able to speak openly about my feelings when angry or worried. 1 2 3 4 5

16. I have regular conversations with the people I live with about domestic problems, e.g., chores and money. 1 2 3 4 5

17. I do something for fun at least once a week. 1 2 3 4 5

18. I am able to organize my time effectively. 1 2 3 4 5

19. I drink fewer than three cups of coffee (or other caffeine-rich drinks) a day. 1 2 3 4 5

20. I take some quiet time for myself during the day. 1 2 3 4 5

To get your score, add up the figures and subtract 20. A score below 10 indicates excellent resistance to stress. A score over 30 indicates some vulnerability. You are seriously vulnerable if your score is over 50.

You can make yourself less vulnerable by reviewing the items on which you scored three or higher and try to modify them. Notice that nearly all of the items describe situations and behaviours over which you have a great deal of control. Concentrate first on those that are easiest to change – for example, eating a hot, balanced meal daily and having fun at least once a week – before tackling those items that seem difficult.

The first problem that anyone needing to manage stress and the stress response must know is that most of us don't even recognize when we are going into stages two and three of the General Adaptation Syndrome – the chronic phases of stress – until it is too late. The failure to recognize the presence of stressors and the physical and psychological effects of the stress response leads to *burnout*. Burnout is now an old fashioned term. The updated term is *vital exhaustion*. Vital exhaustion will eventually set in if the reward/effort ratio is too low. In other words, if your reward is low, for an effort that is high on the long term, eventually that will lead to disease. This becomes a problem if burnout is the first clue that you notice. Its already way too late. Cumulative or chronic minor stressors are particularly insidious because the mind adjusts for them and habituates. We cease to remain conscious that they are there, yet they continue to affect the hormonal and immune systems. This is adaptive on the short run but maladaptive in the long run. We could describe this as a state of anaesthesia or mind-numbing. Even bodily anaesthesia can occur with repeated stressors – ask any long distance runner. Humorously, we could refer to a doctor who puts you in touch with how you really feel, one who makes you aware of your dissociated pain and stressors, as a *de-anaesthetist*. What a subspecialty that would be!

To use our fire metaphor, if you don't realize that the building is on fire until the floor beneath your feet starts to burn, its too late; too late to put it out; too late to save the house; too late to save your possessions.

Please recall the table of physical, emotional and cognitive symptoms of acute stress. These are not likely to be missed. Chronic stress is more insidious and more difficult to recognize, precisely because one can get used to it. We see a similar situation in people with chronic pain – thay may stop complaining but are still affected.

The list below is taken from a popular stress management project.[32] It suggests that if the profile fits, you are experiencing a lot of chronic stress. Even if you do not recognize that you are stressed, others might be aware that you are in quiet distress. Your own brain may be blind to your suffering but chances are that your neighbor is not. The following are early warning signs that you, a loved one, or someone who reports to you is suffering from vital exhaustion (burnout):

- Chronic fatigue – exhaustion, tiredness, a sense of being physically run down
- Anger at those making demands
- Self-criticism for putting up with the demands
- Cynicism, negativity and irritability
- A sense of being besieged
- Exploding easily at seemingly inconsequential things
- Frequent headaches and gastrointestinal disturbances

- Weight loss or gain
- Sleeplessness and depression
- Shortness of breath
- Suspiciousness
- Feelings of helplessness
- Increased degree of risk taking
- Increased drug or alcohol use

The above profile fits the definition of vital exhaustion. I hope that you will be able to control stressors and your body's stress response before it gets to this point. At all costs, if you wish to remain healthy, you must avoid getting to the later stage. Stage three of the General Adaptation Syndrome would certainly indicate the chronic presence of stressors and an associated unmitigated stress response.

Part three – understanding and managing your personal response and doing what works for you

Remember what happens when the body is exposed to a stress trigger? You experience a set of immediate physical, emotional and cognitive changes. If the stressor is mild and short-lived, you quickly go back to normal. If the stressor is more significant and lasts longer, the stress circuits remain activated and the symptoms remain. Recall again the list of common symptoms that people exhibit when they are exposed to stressors. Not everyone experiences all of the symptoms as each person has an individual metabolism and body structure, but everyone will experience some of them. Your first job in the stress management coordinator's role is to identify the set of physical, emotional and cognitive symptoms that characterize your response to exposure to stressors. Get to know yourself. Learning to monitor the occurrence and the severity of these symptoms is fundamental to controlling your stress circuitry.

It is impossible to give a universal formula for managing the stress response. The following section describes effective *stress busters* and strategies to manage the stress response. You should master some of these and develop a repertoire that once learned and practised, will come naturally.

Breath is life

Life flows into us through our breath. When we are focused, centred and present, our breathing is naturally deep, slow and rhythmic. When we are angry, upset, afraid, distracted, our breathing becomes shallow, fast and irregular. Normally, we use only a small portion of our lung capacity. Like balloons, our lungs can expand to hold a far greater amount of air through proper breathing than we normally hold. The deeper and fuller we breathe, the more oxygen circulates into our system to revitalize and energize the body's tissues, organs, and major metabolic systems. Deep and full breathing brings necessary restoration and rejuvenation – and a sense of calm.

Deep breathing practice

Sit in a comfortable position on the floor, sofa or a chair. Now, lift the shoulders up and back, expanding the chest area, up, open and forward. (You can also imagine pulling your shoulder blades together and downward as you sit up straight and open your chest cavity). Now, with your chin and head level, close your mouth and begin to take long deep breaths through your nose (both nostrils if possible). As you continue deeply inhaling and exhaling, become aware of your abdomen (belly) area. Now as you exhale, gently and firmly squeeze or push the navel point (belly button) backwards towards your spine. Then, as you inhale, release the backward push from the navel and allow the belly muscles to relax forward so the lungs now fills with air as you continue inhaling. Continue in this way for one to three minutes. If you get dizzy, slow down.

You may experience lightheadedness. This is normal and is due to blowing off the carbon dioxide and increasing the oxygen concentration in the lungs, blood stream and brain. As you continue to do your breathing practice, you will experience a deep state of well-being. When the breath is controlled, the mind's anxiety is controlled. Conversely, holding your breath will give you an experience of anxiety. Anxiety causes improper breathing and even breatholding. Deep breathing practice will counteract that.

While the above method is probably the most effective way to break the stress response, there are many other behaviours which may be very useful. Learn a small repertoire of stress busters from the table below and practice them every day.

Positive stress response behaviours

Lie on your back, knees bent and touching and do the breathing exercise as above.

Stretch your neck: stand with feet apart, well balanced, and slowly, gently rotate your head and neck for thirty seconds and then reverse direction for thirty seconds.

Stand comfortably and shrug your shoulders vigorously five times.

Express your distress. Write or talk about your stress to someone who can listen.

Take a short walk.

Learn the *Relaxation Response* as outlined by Dr. Herbert Benson (available on-line and at bookstores).

Learn *Focusing* as described by Dr. Eugene Gendlin (available on-line and at bookstores).

Get some quiet time – retreat to some space where you must be alone for a short time.

Listen to relaxing music.

Use visual imagery: sit quietly with eyes closed and visualize different positive ways of behaving under a difficult circumstance.

Have a massage.

Take a bath or shower.

Have a cup of tea.

Read some inspirational or poetic works.

Have a power nap.

Meditate. Use positive, pleasant visual imagery.

Have a brief exercise or stretching routine.

Talk to a friend or colleague.

Use humour and funny entertainment – laugh.

Laughter is one of the best cures for stress. Why? Because laughing reduces the levels of dopamine, adrenaline and cortisol that are involved in the stress response. In short, laughter turns off the stress mechanism. According to Dr.Michael Miller, Director of Preventive Cardiology at University of Maryland Medical Center, laughter is a good stress releaser and people who laugh recover faster from a heart attack. Laughter is a behaviour that can be learned and it has been proposed as part of the treatment after a heart attack.[33]

There are many more stress busters. The main point is that you change the stressful situation. Take control. The loss of autonomy itself, being a victim, is the most stressful. That is why Robert Sapolsky said that the most stressful place on earth was the typical nursing home – because of loss of autonomy.[34]

Music, imagery and touch (MIT) therapies

Research has shown that some of Mozart's sonatas (K44, as noted earlier), and music of that type, can alter brain waves. The phenomenon of entrainment, brain waves and heart rate getting in step with a relaxing tempo, is experienced as deeply relaxing. Beware, not all music is relaxing. The *Mozart Effect* reduces anxiety. It had been used for Relaxation Therapy and even in the treatment of epilepsy. Different parts of the brain respond to musical qualities. The left cortex processes and is relaxed by a particular rhythm and pitch. The right cortex responds to timbre and melody.

The practice of meditation goes back to at least 1500 BC on the Indian subcontinent and 7th century BC in China. The World Parliament of Religions, held in Chicago in 1893 increased North American awareness of meditative practices in the context of religion. Today, meditation is a category on the National Library of Medicine research list. The physiology of stress and meditation has been receiving increased scientific attention because these practices are now known to affect neurotransmitters, neuropeptides and brain waves. Meditation has been shown to change brain waves toward a relaxed alpha state or even to deeper states of relaxation. Newer research suggest that meditators improve their overall brain functions when measured by tests for attention, concentration and executive functions. To use meditation techniques effectively, a teacher has to be involved. Although some people can use tapes and self teaching exercises, a person trained in guiding these sessions can save time. Meditative practice is a skill that usually has to be learned. Yoga can induce meditative effects. If you are not interested in a walking or jogging routine, consider yoga classes. Preliminary research by Dr. George Brainard at Jefferson Medical College in Philadelphia, has shown that yoga lowered the stress hormone cortisol level even in healthy and inexperienced practitioners.[35]

Massage is a touch based therapy and it works because it reduces cortisol levels and physiological changes associated with the stress response. For many people these modalities offer stress reduction. The only problem is that amateurish dabbling with the above may not be effective.

MIT therapies tend to be time consuming and expensive because well trained and experienced therapists are required.

Optimism itself is a powerful protective factor against being stressed. There are many studies to show that having a sense of meaning and optimism are associated with protection from disease and premature death. Physicians and their patients who ignore the results of well conducted studies that conclusively show that emotional status affects health, are not practising scientific evidence based medicine.

We have a delicate clockwork of biological rhythms that determine hormone and neurotransmitter undulations. Many stressors throw a monkeywrench into this intricate machinery. Almost any stressor can disrupt our circadian rhythm and can precipitate sleep and mood disorders. Drugs, toxins, irregular meals, traveling and group pressures are just a few common pathways to the disruption of equanimity. Stay with your usual routines and rhythms as much as possible. This does not mean that you have to become rigid and inflexible. Sleep time, meal times and work time are examples of routines that are better to be left undisturbed. The stress hormone is a bad one. In the next chapter we will learn about some more friendly hormones.

THE LEAST YOU NEED TO KNOW

- The effects of stress are bad for the brain; the greater the stress and the longer it lasts, the worse it is.
- Stress effects are cumulative and stress elevates stress hormone/cortisol levels.
- Symptoms and signs of stress can be recognized.
- The effects of stress should be acknowledged when present.
- There are high risk groups – this may mean you!
- If you get used to stress symptoms that does not mean you are not stressed.
- You can learn coping skills to avoid being stressed.
- It is always important to identify the causes of stress effects so that you do not merely treat the effects.
- Simple stress busters such as breathing and meditation can help in coping with the effects of stress.

START NOW!

- Learn about stress – read *Always Change a Losing Game*, by Dr. David Posen.
- Cultivate a sense of optimism – it is the most protective against stress.
- Recognize signs and symptoms – do the tests in this chapter.
- Take the time to learn quick stress busters.
- If you cannot get a handle on your stress response, seek professional help – treatment exists.

www.HealthyBrain.org

▬▬▬▬▬▬▬▬▬▬

CHAPTER 13

HORMONES

It would indeed be rash for the mere pathologist to venture forth on the uncharted sea of the endocrines, strewn as it is with the wrecks of shattered hypotheses, where even the most wary mariner may easily lose his way as he seeks to steer his bark amid the glandular temptations whose siren voices have proved the downfall of many who have gone before.

– Dr. William Boyd[1]

You may wonder why I include hormones, or better said hormone balance, as one of the Eight Pillars of Brain Health. There are actually three very good reasons for this choice. First, the brain is exquisitely sensitive to hormones and hormone-like substances. Hormones have profound effects on the brain and can influence our behaviours, moods and even the functioning of our immune systems. If your hormonal system is out of balance, your brain will also be out of balance. Second, hormone-related brain disorders are very common. They are also commonly misunderstood but a competent medical doctor can identify these conditions and treat them. Understanding the basics of brain-hormone relationships will allow you to attain a better level of health. Third, hormone replacement therapies are becoming increasingly common and more and more people are turning to them as a way of fighting the effects of aging. Unfortunately, it is a fight that people inevitably lose and in the battle they may inadvertently impair their brain function or initiate some other bodily damage (e.g. increase risk of cancer). It is my duty to warn you about some of these questionable practices. In short, I put this material in the book because you need to know about it, if only to avoid things that could potentially be harmful to you.

What are hormones?

The answer is simple. Hormones are chemical messengers that are generated in the body and/or the brain which communicate with target organs. The system of glands that produces hormones is called the endocrine system. Allow me to introduce you to *psychoneuroendocrinology*, the second longest word in the English language. This rather overblown word describes the study of hormones and how they affect the brain and the nervous system. Psychoneuroendocrinology is in its infancy but is already becoming an important modality for understanding and treating brain disorders. (This is the last time I will use the word in this chapter – it is too much of a tongue twister.)

It is interesting to note that the word *hormone* comes from the Greek *hormaein* which means, to excite, arouse or set in motion. This actually gives a very good clue as to what hormones do – they set things in motion. Most hormones cause something to happen such as sensation of appetite, sleep induction, feeling of well-being or even tissue or organ growth. A word that is based on action and motion seems most appropriate.

How do hormones work?

Let's start with a simple model. Hormones are produced and secreted by an endocrine gland. Once they leave the gland, they are carried through the bloodstream to another organ or to some tissue. When they arrive they will have a very specific effect. The hormone may 'turn on' an organ or biochemical process or change the way it is functioning. Most hormones are *anabolic* which means that they play an important role in the building of tissue. Almost everyone has heard of 'anabolic steroids', the substances that athletes and bodybuilders ingest to improve muscle growth. Those substances are anabolic steroidal hormones. They are purified or synthetic versions of the naturally occurring hormones that our body uses when growing new tissue or healing wounds. Bodybuilders use them because it 'turns on' the growth process and allows for rapid muscle tissue development. That has its dangers.

In fact, naturally occurring hormones are so potent in supporting organs, including brain tissues, that they have been referred to as a 'fountain of youth'. Why? Some people believe that if we can control our hormone levels we can reduce such effects of aging as wrinkled skin, loss of muscle mass and increased fatty tissue – in short, we will be forever young. I believe that is a dangerous misconception. The endocrine system is very complex and hormone replacement, especially the commercial kind, oversimplifies the whole issue. As you read earlier, I posited that the Fountain of Youth is physical exercise – fitness. Fitness produces natural hormone balance and reduces weight gain; therefore, I keep my belief. Physical fitness is the closest we can come to *The Fountain of Youth*! Staying healthy is not an effortless downhill slide using pills.

Interesting hormone features

- Hormones act in very small concentrations; they are measured in parts per million.
- Hormones do not act in isolation. They depend on each other and can modify each other's actions. They may have slightly overlapping additive or subtractive effects.
- They act rhythmically and are released in a pulsatile fashion, in small packets, throughout the day or night, governed by various body rhythms such as the day/night, lunar and fixed innate cycles.
- There are many variations of hormone molecules with similar actions. Synthetic and plant derived substances and animal organ extracts/excretions have hormonal actions but may not contain molecules identical to those found in humans.

When you look at the whole picture, the endocrine system is amazingly complex and because of this, tinkering with it is a risky business. The molecules are so potent they can be harmful in larger than natural concentrations and in ill-timed exposures. Furthermore, entrepreneurs may make exaggerated or even false claims. Overzealous and artless hormone replacement usually results in the 'too much – too late' scenario. Taking yet another pill to gain control of the body has a magical quality and precludes effort. Taking large doses of synthetic hormones in patterns that bear no resemblance to natural secretion patterns and that are not balanced with other systems, is a recipe for disaster.

Hormone hazards

- Hormone abuse for body-building
- Anti-aging hormone treatments
- Overzealous treatment by physicians – iatrogenesis
- Too much too late
- Side effects
- Accumulation of hormone impostor chemicals from foods
- Pesticides

Hormones and aging

As we grow older, a gradual decrease in anabolic hormones (the ones that play an important role in building tissues) occurs after menopause. The result is that most people who experience this will find they have changes in energy level, strength and stamina. Most people don't find this to be either physically or psychologically pleasing and they often seek to stop or reverse the process. That will usually involve hormone replacement therapy. In some cases it is absolutely necessary. In such cases, hormone replacement should be performed in an evidence based, medically proven manner. Diagnostic criteria for an illness, a syndrome, must be met and replacement has to be monitored closely. That means follow-up by a doctor, possibly an endocrinologist, for a long time.

Medical management of hormone disturbances has a long and distinguished (if not disturbing) history. In 1849 Dr. Berthold, inserted the testicles from a rooster into a hen and noticed an amazing thing: the hen soon began to crow like a roster, something that never occurs in nature.[1.1] This led to much speculation about the role of glands and hormones and excited the medical community about the possibility of artificially changing characteristics using hormones. Shortly after Berthold, Dr. Cushing, a neurosurgeon and then Dr. Addison identified disorders that were directly caused by hormonal imbalances. Another important breakthrough in the history of hormones was Dr. Asher's discovery that a lack of thyroid hormone caused severe psychiatric illness.[2]

The troubles begin when people, for whatever reason, ignore the medical guidelines and try to use hormones to fight the natural aging process. When that happens, proper medical treatment disappears and the cult of 'anti-aging medicine' emerges.

Hormones and the fight against aging

The use of hormone application to fight off natural aging processes, to promote sexual performance or to substitute for healthy living, has a long and not too distinguished, although somewhat entertaining, history. While medical science was seeking to understand the nature and actions of hormones, another group of so-called researchers started to look at ways hormone extracts and preparations could be used to improve physical and sexual function. Much of this early work focused on the use of ground up, or otherwise rendered, animal glands, notably testicles, to rejuvenate older men, although women were targeted as well. Over the next hundred years, new preparations, tonics and cures appeared in the press with staggering regularity. The rich turned to expensive spa treatments, usually in European countries. The less affluent turned to the mail order catalogue or newspapers where hormone treatments were shamelessly promoted.

High hopes for hormones

In the late 1800s Professor Brown Sequard prepared extracts of animal glands that he claimed could be injected into human blood. He claimed that his extracts could produce youth and vitality; he appealed to those who were growing older and, coincidentally, could afford his treatments. The price of 25 injections was $2.50 – a lot of money in 1893. Being a considerate soul, he would also sell a special syringe for injecting his preparations for another $2.50.[3] At this time you could buy a normal house for around $800 and most people made less than that in a year. There are no records of the number of people who suffered serious side effects. Neither is there any record of people who claimed to have been 'cured' by Dr. Sequard's treatments.

Now back to science. It will be useful to know a little bit about the endocrine system. The endocrine system's activity is orchestrated by the hypothalamus, a structure that is located deep in the brain, just above the pituitary gland – very close to the roof of the mouth and near the back of the throat. The cells in the hypothalamus monitor the temperature and chemical composition of blood. When adjustments are needed, the hypothalamic cells send out chemical and nerve signals to the pituitary, which respond by secreting a variety of chemical messengers to stimulate endocrine glands, which in turn, secrete hormones to act on distant parts of the body or the body as a whole. For example, the thyroid hormone (thyroxin) regulates metabolic rate and body temperature. What actually happens is a complex cascade of events, starting with the hypothalamus. The hypothalamus monitors blood composition and temperature directly and is also influenced by higher brain regions including the limbic (emotional) system and cortical areas. It sends out thyrotropin releasing factor (TRF) which stimulates the pituitary gland to secrete thyroid stimulating hormone (TSH). It is the TSH that acts on the thyroid gland to pour thyroxin into the bloodstream to accomplish its task of regulating metabolism. The circuit described above is very reliable and stable, as are other endocrine pathways. They usually operate flawlessly for many decades but aging and other risk factors accumulate and the signaling can go out of whack. The endocrine system has its own vulnerabilities – it is not to be messed with. Look at the following interesting physiological phenomenon:

Beating the heat with a warm drink

Think of the hypothalamus as a thermostat close to the back of the throat. If the throat is warmed up, as with hot tea, the hypothalamic temperature monitors respond to program the body to burn less glucose and produce less heat. That cools the body and leaves you more comfortable. Conversely, sucking a popsicle or putting ice in the throat cools the throat and this prompts the hypothalamus to turn up the metabolism and produce more body heat. Children can get heat stroke this way on hot summer days. A similar phenomenon would occur if you put ice around your thermostat on a hot day. It would cause the heat to turn on in your house and make you unbearably warm. You would not do that so why give your kids popsicles? They will tend to get hotter and want more – to cool off, of course!

That is certainly good for the popsicle industry. Did you ever wonder why people who live in the tropics drink hot tea?

Hormone balance and aging

In both men and women, hormone secretions tend to diminish with age. An exception to this is cortisol which actually tends to increase with age. With a decline in the presence of anabolic (tissue building) hormones, the body, as it ages, slowly slips into what is called *negative protein balance*. For all of us, this means that after menopause (women) and andropause (men) years, we lose more protein, every day, than we gain. The effect of that imbalance is that we actually start to shrink. The shrinking effect isn't dramatic, but it does mean that our body composition and structure changes. If you don't believe this, measure yourself as you age and you will find that you are getting shorter; don't feel bad, so is everyone else. As people reach extreme old age they become increasingly frail as they lose muscle and tissue mass. Eventually, the negative protein balance leads to organ failure and a natural death. This is completely different from unnatural or premature death caused by a specific killer disease. The endocrine system produces a large number of hormones and hormone-like substances. You don't need to know all of them, thank goodness, but there are some that are common and it is worth recognizing their effects:

Hormone	Effect on body and brain
Androgen (testosterone)	Male sexual characteristics and behaviours
Antidiuretic hormone (vasopressin)	Blood pressure regulation; learning
Calciferol (vitamin D)	Cell differentiation, cell growth, neurogenesis, mood
Estrogen (estradiol)	Female sexual characteristics and behaviours
Growth hormone (somatotropin)	Body growth (muscle and connective tissue)
Insulin	Regulates blood sugar entry into cells; lowers blood sugar
Leptin	Curbs appetite
Melatonin	Core body temperature, biological rhythms, sleep
Oxytocin	Parturition (labour), bonding, trust
Parathyroid hormone	Bone metabolism, anxiety, mood

Progesterone	Stimulates breast tissue and secretion of milk suppresses ovulation, mood effects
Thyroid hormone (thyroxin)	Metabolic rate, body temperature

Let us discuss each hormonal system, in alphabetical order, not in the order of importance, and explore how each one links to brain health.

Androgens

Androgens are male sex hormones that are produced in the gonads and are responsible for typical male sexual characteristics and behaviour. The most well known androgen is testosterone. Testosterone has a number of effects on behaviour and mood including a potent effect on sex drive and arousal. It also has an anabolic affect that enhances metabolic processes in muscle, bones, bone marrow and the immune system, and it also improves or enhances cognition and mood. It does all that in males and females.

Testosterone has had a bad rap because of its commonly perceived association with sexually predatory and violent behaviours. In fact, testosterone appears to be able to pump up aggressive drives to the point where the cerebral cortex cannot inhibit and modulate the erupting behaviour. This may explain why young men have higher rates of impulsivity and violence than other age groups – their testosterone levels are at their lifetime highest peaks. There is also clear evidence that some people who abuse testosterone to enhance drive, muscle building and athletic performance are susceptible to a number of side effects including brain effects manifested by aggressive behaviour, ill temper and sexual impropriety. The term 'roid rage' refers to a syndrome of paranoid homicidal rage precipitated in men who through abuse, have unnaturally high levels of androgenic steroid hormones. 'Road rage', another modern phenomenon, not necessarily related to androgen abuse, also has its roots in irritability and hair trigger temper characteristic of testosterone-primed men.

Andropause is a modern male issue. Androgen deficiency in the aging male (ADAM syndrome), or andropause, is a condition experienced by men during late middle age and is associated with a decline in total testosterone levels. It is estimated to affect 20% of men over the age of sixty.[4] The symptoms of andropause include a loss of libido, erectile dysfunction, weakness, fatigue, lethargy, insomnia, mood disorders, less motivation and sometimes depression. Andropause is a confluence of disquieting symptoms and signs that overlap with those of depression. The latter is understandable as researchers have established that testosterone produces an antidepressant effect. Men suffering from this syndrome may be referred to a psychiatrist for seeming 'depression' but antidepressants and psychotherapy do not help them – they need testosterone replacement therapy. Andropausal males may also experience osteoporosis and loss of bone density but not as rapidly and profoundly as women.[5] Current medical thinking upholds that it is inappropriate to accept the idea that older men with symptoms of low testosterone levels should live with that. If there are no contraindications, treatment should be available.[6]

Antidiuretic hormone

Antidiuretic hormone, also known as vasopressin, has several important effects. It is a powerful constrictor of blood vessels which causes elevation of blood pressure and it regulates the composition of the blood, particularly the sodium content. In addition to this it has some neurotransmitter activity in the brain that affects memory, learning and passive-avoidant behaviour. Vasopressin has been shown to improve memory in human subjects and in cognitive impairment, but not specifically in AD and generally not in females. Animal studies show that it can alter learning of avoidant behaviour. The vasopressin receptor has also been associated with sociability and faithfulness among prairie voles. Voles are small fieldmice and the males tend to be faithful to their mates for life. Males share in the care of offspring, but some males are promiscuous and inattentive fathers. One molecule, a vasopressin receptor, makes the difference. Males with prolific vasopressin receptors in specific brain areas, including the amygdala and hypothalamus, spent more time with their mates, were more likely to remain with their mates when new females were placed near them and spent more time grooming their pups than the males with low levels of vasopressin receptors. We do not know the implications for humans.[7]

Calciferol (vitamin D)

Vitamin D is not a vitamin, it's a hormone.[8] It has been called the 'sunshine hormone' but it would be better named as the *hormone of cell differentiation*. It is synthesized by the skin when ultraviolet rays from the the sun shine on it and it travels in the bloodstream to all organs of the body including the brain, where it promotes cells to grow toward their final structure and function. It does that by regulating genetic programs in the nucleus of every cell. Low vitamin D levels are common in people living north of the 42nd parallel (north of the lattitude of Boston), particularly in young people, the elderly and those with darker skin. Vitamin D is better known for its role in calcium absorption and bone development. Chronic insufficiency has some dangers other than weak bones. Optimal cell differentiation is necessary not only for bone formation but also for proper development and function of muscle cells, immune system cells, nerve cells and in particular hippocampal brain cells. Vitamin D is believed to have important effects on brain cells promoting division and growth (neurogenesis) through the action of molecules called *brain derived neurotrophic factors* (BDNF).

Vitamin D also helps parathyroid function by suppressing its overactivity. It is not surprising that vitamin D insufficiency has been linked with depressive illness, in particular winter depression, memory problems, low energy, muscle aches and pains, poor balance and some serious brain diseases (multiple sclerosis and some brain tumours). Animal studies show that deficiency during gestation results in offspring with anatomical brain abnormalities.[9] We do not know if that happens in humans. Unfortunately, this substance has not been studied very thoroughly, thus deserving the nickname, 'the neglected hormone'. Of course, just because we failed to gather scientific evidence about the actions of this substance does not mean that important actions do not exist.

Recently, newspapers have been full of marvelous new findings about vitamin D. This has been in connection with its anti-cancer properties (recall that it is a hormone of cell differentiation and cancer cells tend to be undifferentiated, primitive invasive cell lines), not particularly relevant to brain health. The studies I was describing pertained to vitamin D's neurosteroidal effects on the brain.[10] Of course, it has anti-cancer effects in the brain too. Stay tuned – Health Canada recently, revised its guidelines on vitamin D supplementation. We are now advised to supplement vitamin D 400 international units (IU) per person every day, in addition to the food we eat. This includes children. The Canadian Cancer Society increased this limit to 1000 IU per day.

There is a false sense of danger about vitamin D toxicity. That's a result of the fact that most often, vitamin D preparations come with vitamin A in much higher than physiological doses. In other words, if you take 'vitamin D' as an over the counter generic preparation you may be getting a huge and, on the long run, dangerous doses of vitamin A. When you supplement vitamin D you must be sure that it is just vitamin D and nothing else added with it. Even 'natural' liver oils, when yielding the proper amounts of vitamin D have too much vitamin A. Stick to pure vitamin D as a supplement and do not go over 1,000 IU per day unless you are advised to do so by your physician or an endocrinologist. Of course this advice applies to those north of the 42nd parallel; people in the south, unless shielded from the sun, need not supplement. It is worth repeating that people with darker skin need more Vitamin D in the northern latitudes. It is estimated that the unclad human skin will generate 10,000 IU of vitamin D per sunny day.[11]

What is the best way to optimize vitamin D intake? You could go out in the the sun because the sunshine on your skin stimulates vitamin D production. Dermatologists tell us that too much sun is dangerous, especially as we age, because it causes skin cancer. You could drink more milk because the government puts vitamin D in our milk to prevent rickets. However, some people are lactose intolerant, or gain weight when they drink milk. You could take a vitamin D preparation under your doctor's advice. That is my favourite solution – no skin cancer risk, no flatulence and diarrhea from lactose and no weight gain from the fat in milk. By the way, vitamin D is fat soluble which means that it travels in fatty substances (fish oils, milk), but defatted foods have less fat and less vitamin D.

Estrogen (estradiol)

Estrogen is a general term for female steroid sex hormones that are secreted by the ovaries and are responsible for typical female sexual physical and behavioural characteristics. Sometimes it is spelled 'oestrogen' because it got its name from the oestrus cycle of mammals, denoting going into 'heat', usually once a year to reproduce their species. Human females do not go into 'heat' the same way, but they do experience shifts in monthly estrogen levels, fertility and mood based on their monthly cycle. Estrogen has interesting and wide-ranging effects. It can affect circulation by dilating blood vessels that account for the hot flushes women experience during hormonal fluctuations accompanying menopause. Low estrogen levels found in menopause are associated with impairment of thyroid response to thyroid stimulating hormone. This is a type of subclinical hypothyroidism which can be reversed by estrogen replacement therapy (ERT).[12] Interesting data from animal studies suggest that estrogen promotes brain cell growth and synaptic formation in the hippocampal and cerebral cortex, brain regions fundamental to cognitive functions.[13] Experimentally it has been shown that ERT in post-menopausal women improves blood flow to the brain and reduces age-related memory problems.[14] PET scan studies showed better perfusion in hippocampal and temporal lobe areas.[15] Findings suggested that ERT may lower susceptibility to Alzheimer's disease.[16] Unfortunately, there is a trade-off. More about this later.

Estrogen also stimulates multiple neurotransmitters, enhances cerebral blood flow and acts as an antioxidant, thus reducing the oxidative damage and subsequent deposition of the neurotoxic amyloid beta protein seen in AD. Although animal studies were promising, human studies of estrogen effects on the brain are less conclusive; more recent studies suggest that it increases the risk of stroke in women.[16]

Traditionally, estrogen was prescribed to protect women against cardiovascular disease, bone disease and even cognitive impairment but this practice is now questioned. Although the anabolic effects of estrogen do help bones and brain, the risk of cerebrovascular disease with cognitive impairment and strokes and an increase (rather than decrease) in heart disease risk outweigh the anabolic benefits.[17] In a healthy woman, estrogen and progesterone are in balance. Shifts in estrogen levels can cause unpleasant mood swings, problems with concentration and memory, irritability, insomnia and physical symptoms (sweating, flushes, swelling). High estrogen levels decrease interest in sexual activity.

A significant study, the Women's Health Initiative Memory Study, alerted Canadian health care professionals that combination hormone replacement therapy may increase a woman's risk of neurocognitive impairment, most likely through cerebrovascular disease from the accumulation of small subclinical strokes.[18] Instead of long-term estrogen replacement, we are now advised to use the lowest effective dose and for a limited rather than prolonged period.[19] Estrogens also increase cancer risk. Long term use of estrogen postmenopausally increases breast cancer risk. Given alone, it increases risk of uterine cancer, an effect which appears to be blocked by progesterone.[20] Breast cancer risk is increased by 40% in women taking both hormones.[21]

Another issue about estrogen replacement is where the estrogen actually comes from. Historically, much of the estrogen used for hormone replacement was isolated from the urine of pregnant mares. It so happens that the estrogen compounds coming from pregnant horses are not identical to the human molecule of estrogen which is *estradiol*. Although not thoroughly studied, it appears that low doses of estradiol, merely to replace diminished levels, may be safe.[22]

To complicate things further, we know that there are many estrogen-like molecules from plant sources and also from substances introduced by mankind into the environment. The insecticide DDT is very similar in structure to estradiol and a synthetic female hormone drug with estrogen like effects called diethylstilbestrol (DES). DDT has worked its way into our food chain and although its use has been abandoned in developed nations, it, along with other estrogen-like imposter chemicals, acts on estrogen receptors in humans and animals. Depending on age, sex and hormonal status, the same dose of such estrogen mimicing compounds can have wildly different effects.[23]

Growth hormone (somatotropin)

Growth hormone as the name implies causes growth. You can tell whether your growth hormone levels have dropped; if your chest seems to have dropped to your abdomen your growth hormone has dropped. This is a hormone that comes from the pituitary gland that promotes body growth, fat mobilization and inhibition of glucose utilization. When excessive, it can cause diabetes and some other nasty side effects such as heart disease, bone deformities (acromegaly) and brain tumours (meningiomas). Milder side effects include carpal tunnel syndrome, swelling from fluid retention and joint pains. In Canada, we don't treat patients with age-related decline of growth hormone, although treatment is available for adults with clearly documented growth hormone deficiency. Such documentation requires testing by an endocrinologist. Growth hormone replacement in deficient adults may confer a number of metabolic improvements and a sense of well-being.[24] It has been used post operatively to improve recovery.[25]

Growth hormone does decrease with age and entrepreneurs are quick to point this out. They will claim that their products will help you combat the effects of aging. While it is true that growth hormone levels drop as we age, there is no scientific evidence whatsoever that lack of growth hormone itself causes aging or that taking growth hormone slows aging. The evidence that growth hormone grows muscle and gets rid of fat is highly controversial because other effects also kick in. Furthermore, it is against the law to sell it without a prescription, so these products available over the counter or through mail order do not contain any of the hormone. Products that are sold as 'growth hormone releasers' are just amino acids, the building blocks of protein, that are the same as the protein you get in your food. Anything that you eat can be called a growth hormone releaser because all foods raise blood levels of growth hormone slightly and temporarily.[26] When you eat protein, blood levels of growth hormone rise even higher. Growth hormone releaser pills cost much more than food and have not been shown to raise blood levels better than the ordinary foods you eat every day. Exercise can also be called a growth hormone releaser because every time that you exercise, blood levels of this hormone rise.[27] Exercise raises growth hormone levels more and longer than eating does. Recent research shows that growth hormone levels are lowered by having lots of fat stored in your belly. Anti-aging claims are fanciful because we have no scientific way of measuring aging. The commonly used tests to measure aging actually measure fitness. No hormone will produce fitness (see Chapter 9, Physical Exercise). To reduce the effects of aging and improve your performance on all medical tests of aging, start an exercise program. If you want to gain muscle and lose fat, reduce your intake of refined carbohydrates and fatty foods; eat plenty of the foods that come from plants (fruits, vegetables, whole grains, beans and other seeds) and maintain a regular, vigorous exercise program. Beware of anti-aging quackery. It cannot be denied that growth hormone has a number of intriguing effects that may someday be exploited for the management of clinical conditions. In the future, it may have utility in improving healing after some types of surgery or injury.[28] More research is needed. We are not there yet. Our knowledge has to grow before we grow from growth hormone!

Insulin

The most common endocrine disorder is diabetes. Many people don't think of diabetes as an endocrine disorder but insulin is a true hormone and resistance to it or lack of it will produce the diabetes syndrome. Diabetes has been called 'starvation in the midst of plenty' because it is accompanied by high blood sugar but this sugar cannot get into the cells where it is needed as a source of energy. Insulin is the 'hormone of starvation'. The inability of sugar from the blood to enter into cells gives rise to high blood sugar levels that interact with the fat in the blood to produce blood vessel disease and clogging of arteries. That is why people with diabetes have poor circulation and are prone to peripheral nerve damage. Other complications include gangrene, coronary arterial heart disease, cerebrovascular insufficiency and strokes.

Let's look at diabetes a little more closely. Diabetes affects about 17% of people in Canada (and USA), mostly in the over-forty age bracket.[29] This is a huge public health problem. Diabetes is a chronic and essentially incurable disease. It is the fourth cause of death by disease with high rates of disability, morbidity and mortality because of life-threatening complications. Many people don't realize that diabetes is an independent risk factor for cerebrovascular disese and major neurocognitive impairment. Cerebrovascular neurocognitive impairment (we are still avoiding the term dementia) is a serious, life-threatening brain disease, second in frequency to AD. There has been an estimated fivefold increase in the prevalence of diabetes over the last thirty-five years (prior to 2007).[30] This means that from about 1970-2005 the numbers have risen to 8 million in the USA and about 1.5 million in Canada.[31] The disease remains under-diagnosed. One case goes undiagnosed for every two cases of diabetes (type II) that are diagnosed and treated. It consumes about one-seventh of health care dollars. Diabetes is associated with depression and depression is a brain disorder. The prevalence of depression in diabetes is about three times that of the general population.[32] In the general population the prevalence of depression is about 7% but in the diabetic population it's about 25%.[33] At any given moment, one in five diabetics suffer from depression. The lifetime prevalence data indicates that one in three diabetics will have suffered depression.[34] Interestingly, treatment of depression to full remission also improves diabetes.[35]

Another insidious syndrome related to diabetes is Syndrome X. This syndrome is also known as *metabolic syndrome, insulin resistance syndrome* or Reaven's syndrome.[36] This syndrome is associated with advancing age, primarily affecting people in midlife. It is also characterized by low activity sedentary lifestyle, smoking and a family history of diabetes. These people, although not necessarily frankly diabetic, tend to have high insulin levels and high blood sugar levels. They may have impaired glucose tolerance which would indicate that they are slipping into a true diabetic illness. It is also characterized by central abdominal obesity and high body mass index (BMI) and elevated waist-hip ratio. Unrecognized and untreated, this syndrome may progress to CHAOS syndrome. CHAOS stands for coronary heart disease, hypertension, adult diabetes (type II), obesity and stroke. People afflicted with Syndrome X often have chronically elevated cortisol levels indicating that they are chronically stressed.[37]

Leptin

The word *leptin* comes from the Greek *leptose* meaning thin. It is a 'hormone of weight loss' secreted by fat cells acting on specific cells in the hypothalamus to curb appetite and to increase energy expenditure as body fat stores increase. Leptin levels are 40% higher in women, and show a further 50% rise just before menarche before returning to baseline levels.[38] Levels are lowered by fasting and increased by inflammation. Human genes encoding for leptin and the leptin receptors that mediate the weight reducing actions of leptin have been identified. Laboratory mice that have mutations on this gene become hugely obese, diabetic and infertile.[39] Administering leptin to these mice improves glucose tolerance and increases physical activity that reduces their body weight by as much as 30% and restores fertility.[40]

Leptin receptor genes have been found in a small number of morbidly obese humans with abnormal eating behaviour. Dr. Stephen O'Rahilly, an endocrinologist at the University of Cambridge found a leptin-deficient child who weighed 94 kg (207 lb.) at age nine.[41] Fortunately, only a tiny percentage of obese people have leptin deficiency. The majority of obese persons do not show such mutations and have normal or elevated circulating levels of leptin which means that their fat cells are trying to shut down their appetite.[42] They eat too much, exercise too little and cannot blame their genes.

Melatonin

The 17th-century French philosopher René Descartes considered the pineal gland to be the seat of the soul. Located at the roof of the posterior third ventricle, it has no direct connections with the brain and is free of the blood-brain barrier that is between the brain, the cerebral circulation and the cerebrospinal fluid. It receives nerve connections only by the sympathetic nervous system and secretes its hormone, melatonin, in response to darkness and low blood sugar. Melatonin, the 'hormone of darkness' is synthesized from serotonin and plays a role in the regulation of chronobiological rhythms, such as sleep-wake cycles. When compared to young controls, the circadian profile of plasma melatonin of old subjects, both healthy and neurocognitively impaired, has been reported to be clearly flattened.[43] The selective impairment of nocturnal melatonin secretion was significantly related to both age and severity of mental impairment.[44]

Data supporting the hormonal dysregulation hypothesis of depression suggest that normal diurnal rhythms are altered or disrupted in acute depression. It appears that mood and neurocognitive function and performance fluctuate depending on the time of day. That has implications for the timing of neuropsychological assessments.

Studies suggest that melatonin exhibits its therapeutic effect for circadian sleep disorders through temperature-lowered induction of the circadian phase shifting rather than by direct sedative effect.[45] This explains why melatonin induces sleep but does not cause cognitive impairment like regular sleeping pills, which depress the nervous system and impair higher brain functions. Pilots are allowed to take melatonin but not hypnotic sleeping pills.

Oxytocin

Until recently it was believed that the only function of oxytocin was to stimulate contractions during childbirth and once a child was born, to stimulate milk ejection on demand by the baby. Recently, it was discovered that there were oxytocin receptors and pathways in the brain and the rest of the nervous system. It is now known that the behavioural effects of this hormone are relevant to psychiatric and psychological states.[46] For decades, oxytocin has been known to have amnestic properties that could play a role in producing cognitive deficits that are routinely observed and reported in both depression and major neurocognitive impairment, especially in females.[47] Preliminary research has documented elevated levels of oxytocin in the hippocampus and temporal lobes of patients with AD.[48]

More recently, scientists have learned that oxytocin has some very peculiar effects on emotional functioning. Swiss researchers at the University of Zürich have discovered that people who inhaled a nasal spray containing oxytocin, were much more willing to part with their money in a 'trust' game designed to measure levels of gullibility under experimental conditions.[49] Concerns have been raised that this finding could be misused to induce trusting behaviours that could then be exploited. It is interesting to note that a hormone that was thought to affect only smooth muscle in the uterus or the nipples of childbearing women could have subtle yet profound emotional effects in all people. These emotional effects, trust in particular, have adaptive value between mother and infant but would certainly backfire in a predatory entrepreneurial world. In some other species, notably voles, this hormone has been shown to regulate behaviours such as pair bonding, maternal care and the ease with which the animal will approach a stranger.[50] It appears that oxytocin is the 'love hormone'. (See also antidiuretic hormone (vasopressin) covered earlier in this chapter.)

Parathyroid hormone (parathormone)

The third most common endocrine disorder is hyperparathyroidism. Parathyroid hormone effects bone metabolism and too much of this hormone increases bone cell turnover leading to elevated calcium in the blood. This condition exists when there is a shortage of vitamin D or if there are tumors producing too much parathyroid hormone. It is more common in northern latitudes. Hyperparathyroidism peaks between ages fifty and seventy and is three times more common in women. In Sweden it affects 3% of postmenopausal women. It is asymptomatic in 50% of cases, meaning that patients do not have associated conditions such as kidney stones or painful bone destruction. Two-thirds of victims have neuropsychiatric symptoms.[51] Such symptoms include fatigue, weakness, depression, cognitive impairment and delirium. Suicidality, homicidal behaviour and psychosis have also been reported. The elderly are more susceptible to blood calcium fluctuations.[52]

Progesterone

Progesterone is the 'pregnancy hormone'. It counteracts estrogen and causes infertility while it prepares and maintains the uterus for pregnancy – it offers natural birth control. Progesterone has significant effects on the brain. It reduces sexual behaviours and has a calming effect. It has been used to reduce sexually aggressive and predatory behaviours in so-afflicted men. It may also be called the 'hormone of expectancy' as it seems to produce a mood state comparable to patience and hopeful waiting. Expectant mothers are truly nice people (as long as they don't have to wait too long). What a nice hormone! Too much of this hormone may cause a depression-like syndrome. Post-partum depression affects about 10% of new mothers. This does not mean that progesterone is the only cause. Thyroid disorder is just as common after pregnancy and cortisol levels may also be very high, sluggishly returning to normal. Nevertheless, it is known that progesterone is a neuroactive steroid that modulates receptors that mediate anxiety.[53]

Prolactin

Prolactin is a hormone that stimulates the secretion of milk and possibly, during pregnancy, breast growth. Too much prolactin is usually a side effect of antipsychotic medications. In women, it may induce a mood syndrome; in men it may cause apathy. Even low doses of antipsychotic medications called neuroleptics can cause this syndrome and the elderly are believed to be particularly susceptible to this disorder.[54] Patients with major neurocognitive impairment who exhibit psychosis and disturbed behaviours remain likely to be treated with antipsychotic medications, and the resulting prolactinemia may lead to an increase in the severity of cognitive impairment.[55] Sometimes high levels of prolactin are caused by a tumor in the pituitary gland. Other medications may cause elevated prolactin levels. These include methyldopa used for treating Parkinson's disease, amphetamines, opiates and cimetidine. Tumors of the pituitary gland are relatively common and make up 10 to 15% of all tumors in the brain. Of those, tumors producing prolactin are the most common.

Thyroid hormone (thyroxine)

Thyroid disorders are common. Hypothyroidism is the second most common endocrine disorder – 3% of the general population is afflicted and it is ten times as common in women than men. It is also ten times more common in older, postmenopausal age groups. 5-10% of postpartum women are afflicted. 10% of hypothyroid conditions are treatment resistant.

Thyroid disease has important implications for optimal brain function and brain health. It has been well known that hypothyroidism can mimic major neurocognitive impairment syndromes and depression. Those conditions could often be cured by the careful administration of thyroid hormone. Some clinicians and researchers wonder whether hypothyroidism can actually cause brain damage.[56] There seems to be a strong association between thyroid conditions and brain disorders including major neurocognitive impairment. 10% of all patients with psychiatric illnesses have hypothyroidism concurrently. Some researchers are concerned whether hypothyroidism can cause AD directly.[57] There are sophisticated arguments but only circumstantial evidence for the following:

- Hypothyroidism indirectly causes AD by increasing vascular risk factors such as type II diabetes and hyperlipidemia.
- Hypothyroidism directly causes AD.
- Early presymptomatic AD pathology causes hypothyroidism.
- Another, possibly autoimmune, factor causes both hypothyroidism and AD.

At present, most clinicians are aware that even a slight increase in the releasing factor for thyroid hormone in the body, TSH, may be a harbinger of hypothyroidism. That is why we measure TSH and assess thyroid status in patients when we suspect depressive illness, AD or other causes of neurocognitive impairment.

We are awaiting research to elucidate the relationship between vascular risk factors and increased risk for AD in patients with hypothyroidism. We need studies, using reliable biochemical tests, on people who have hypothyroidism and AD, to see whether there is a sequence of events suggesting causality. There may also be a relationship between how the immune system functions and endocrine disorders, particularly thyroid disorders. An autoimmune disease is a disorder of the immune system manifested by attacks by the immune system (autoantibodies) on the body itself. We know that this can happen to the thyroid and we suspect that it may happen to the brain. Could AD be an autoimmune disease, like some other endocrine disorders?

Today we have very precise methods for measuring the level of thyroxine in the blood and we can supplement shortages very carefully with tiny amounts of thyroxine. Furthermore, what is amazing is that treatment is done, day in and day out, in your family physician's office with very good outcomes. Before accurate diagnosis and treatment, people with hypothyroidism went mad, the syndrome was called *myxedema madness* and those poor souls were locked up in chains in the mad-houses of antiquity.

Early sources of thyroid hormone came from the dried thyroid glands of animals. Today we use synthetic thyroxin, the human equivalent molecule. Beware of 'natural' sources of thyroid hormone as it is dangerously active and accurate dosing is almost impossible. There have been community-wide outbreaks of hyperthyroidism from thyroid poisoning caused by the consumption of thyroid gland from cattle. In Minnesota, South Dakota and Iowa in 1984 and 1985, thyroid tissue found its way into ground beef. Knowledge of this resulted in the prohibition of gullet trimming in all plants that slaughter cattle and pigs. This syndrome can also happen to anyone who slaughters farm animals or game and trims the gullet to prepare meat for personal consumption.[58]

The future of endocrinology

If hormones are so powerful and can have such great effects on the brain, why don't we use them more often? Historically this is exactly what was done. When a hormone was first discovered there was always great enthusiasm and the public tended to oversubscribe and physicians to overprescribe. It always tended to be a case of 'too much too late' and overdosing was common. Long-term side effects were seldom known in these early enthusiastic phases and have usually been discovered only decades later on. As sophistication with measurement, dosing and the purity of preparations was achieved, proper use of the hormone was finally understood and rational prescribing could be followed. Even today, we do not have easy, inexpensive and accurate ways of monitoring circulating hormone levels. It is possible to obtain blood levels for other hormones besides thyroid but these are more difficult to interpret and are often not used clinically. This will change, of course, as the technology of measuring hormone levels in tissue fluids improves.

At the present time we take blood levels by drawing blood, storing it in appropriate containers, sending it to laboratory doing the testing in batches and reporting the results back to the doctor. Many hormones are carried by protein in the blood, and we are often not getting the accurate level of the pure, free, unbound active form of the hormone. Newer techniques involving checking the saliva for the hormones exist but these have not yet been standardized and are not available to clinical practitioners. This method promises to be much less expensive and accurate enough to allow serial measurements without the complicated collection, storage and shipping problems associated with blood drawing. Taking serial measurements even several times a day will allow the physician to obtain an accurate profile of hormonal activity which then will enable him or her to use the results diagnostically and to guide treatment. More excitement to follow.[59]

THE LEAST YOU NEED TO KNOW

- Vitamin D is a hormone – there are no adequate food sources.
- The three most common endocrine disorders are diabetes II, hypothyroidism, and hyperparathyroidism, in that order.
- Deficiency must be quantified prior to treatment.
- Hormone levels must be adjusted and monitored carefully over a period of time.
- Hormone requirements change with age.
- Hormone requirements change if intercurrent medical illness is involved.
- Being stressed will upset hormone balance and this will also change requirements.
- Consultations from endocrinologists are underutilized.

START NOW!

- If you live north of the 42nd parallel (north of the lattitude of Boston) then supplement with vitamin D.
- Renounce all over-the-counter and mail order hormone preparations, herbal or otherwise.
- Do not assume that hormone replacement is necessary for everyone.
- Insulin resistance (diabetes) must be monitored.
- If you are of menopausal age, male or female, get a hormone panel from a reputable source, an endocrinologist, even if you have to pay for it – this will serve as a screening test and a baseline for the future.
- If you are of menopausal age, whether you are a woman or a man, but you do not have severe symptoms, you should have periodic checkups anyway – at the least have your insulin, thyroid and parathyroid hormone levels checked every year.
- If you are of menopausal age and have symptoms, you must see your physician and you may have to request that you see an endocrinologist.

www.HealthyBrain.org

CHAPTER 14

TREATMENT OF DISEASE

The most frequently used drug in medicine – the doctor himself.
                                        – Dr. Michael Balint 1965

I will discuss the Treatment of Disease in two sections. First we will review common diseases which are major risk factors for the development of brain disease causing neurocognitive impairment. Secondly, we will explore some of the difficulties facing patients and doctors engaged in obtaining and providing appropriate health care.

Section I: Diseases that are risk factors for major neurocognitive decline

Recognition and treatment of risk factors is a true pillar of brain health simply because there are a number of common conditions that are not usually thought of as precursors to early onset brain disease causing neurocognitive impairment. There are two aspects to dealing with these syndromes: early diagnosis and treatment, including follow-up. Sounds simple, yet it is not done well. The former is your doctor's job. The latter is more fuzzy. Who owns the brain? Whose brain should look after yours? Although there is an illusion that 'doctors look after us', they are actually far too busy to do that properly. I own my brain and my brain is ultimately my best health care system. And the same goes for all of us.

Before we go ahead, let's look at a case study of a woman who was proactively involved with staying well. While she didn't intend to promote her brain health, that is, in fact, what she was doing.

Mrs. Smith (pseudonym), was an 82-year-old retired teacher. I was referred to her when I made inquiries about meeting another super-senior. She was living at a local retirement residence, one of the *Independent Living Communities* I visit in the course of my work as a geriatric psychiatrist. Where she lived was the home of her choosing; she has lived there for almost two years. When she found out that I was interested in interviewing her about her lifestyle she treated me to a gourmet dinner in their dining room. While we talked it became clear to me that she was bright, energetic and fully in charge of her self and her environment; she put a high value on *autonomy*. I could not help making some comparisons.

Auguste D. was only 51 years old when she was admitted to the hospital in Frankfurt, Germany in 1901. Her doctor was Dr. Alois Alzheimer, who himself died a few years later, not of brain disease but of cardiovascular disease at the same age of 51, the average life expectancy in those days. She was the first case of documented AD which was considered to be a rare disease of the brain. I recall the words of Elie Metchnikoff, the famous Russian bacteriologist who, in 1903, said, "Old age will be postponed so much that men of 60 to 70 years of age will retain their vigour and will not require assistance". So now we are here, with an 82-year-old woman who works out regularly, participates in social events and enjoys her children and grandchildren. She does this purposefully, with forethought. She consciously valued physical activity and chose her retirement community because they had a well-developed *Wellness and Vitality Program*. It was remarkably similar to *The Healthy Brain Program*©. Emphasis was on physical, emotional, spiritual, vocational, intellectual and social activities. She was stimulated and energized without feeling stressed because all of her activities were voluntary, without pressure or regimentation. That's why she preferred this low stress environment compared to a hospital-like long-term care facility. Physical activity was very important to Mrs. Smith. She exercised in the gym for short period almost every day. That took no great effort because it was integrated into her lifestyle. "It's never too late", she quipped. She enjoyed walking, hiking, bowling, curling, stretching and resistance training. She was self-motivated but also enjoyed the support and motivation from her friends. She said, "You know, if I don't show up for my exercises after a day or two, my friends will miss me and that motivates me". I congratulated her because she gave herself permission to slow down, let someone else look after activities that she found monotonous and unrewarding and she had the courage to explore new activities with her newfound friends including outings to art galleries, concerts, botanical gardens and visits with family members in town. What a contrast to Auguste D. In actual fact, I do not know whether Mrs. Smith had Alzheimer disease-like lesions in her brain; she certainly did not have many signs of it. Nor was she worried about it. Something was protecting her. Like the one-third of the nuns from the Nun Study, who were found to have amyloid deposits characteristic of Alzheimer's plaques, but who have escaped mental deterioration, Mrs. Smith, with or without Alzheimer plaques, had ample brain reserve. She looked well, took care of herself and she enjoyed life. I posit that her lifelong interest in being active, both physically and mentally, was a far more powerful influence in keeping her healthy than her simple genetic endowments. Her brain was leading the way.

Common disorders which parade through the general practitioner's office are not usually thought of as precursors to brain disease and are not taken seriously enough – not by doctors, nor by their patients. The very diagnosis, the name itself, may obscure the issue. The following common ailments (I call this list the *Dirty Dozen*) are listed alphabetically, not in order of importance:

- Alcoholism and substance abuse, including smoking
- Anxiety, stress and depressive disorders
- Blood pressure disorders: hypertension and chronic hypotension
- B-vitamin malabsorption and poor methionine metabolism
- Coagulopathies: increased or decreased tendency for the blood to clot
- Diabetes type II: insulin resistance, high blood sugar, obesity, (abdominal obesity) and the metabolic syndrome (over-nutrition syndrome, Syndrome X)
- Hormonal (endocrine) disorders
- Inflammatory disease of connective tissues (autoimmune disorders)
- Lipid disorders: low 'good cholesterol' and high 'bad cholesterol'
- Periodontal disease (chronic gum infection)
- Poison exposure and ingestion: mercury, lead, POPs, etc.
- Vascular diseases: cerebral, cardiac and peripheral vascular disease and atherosclerosis

Scientific evidence and clinical observations clearly point to the above disorders which, if left undiagnosed and untreated, will lead to accelerated onset of brain degeneration leading to *Major Neurocognitive Impairment* – what we, as neuropsychiatrists, before the new DSM-V classification, used to call 'dementia'.

I will discuss, very briefly, each of the above conditions but even before that, a number of features must be recognized. Each of the *DirtyDozen* conditions share the following:

- They are common – the 'bread and butter' of your GP's practice.
- They are relatively easy to diagnose – any GP should be able to do it.
- Treatments exist; they are usually easily accessible – in most cases free.
- They tend to be chronic, requiring ongoing management – there is no magic bullet to cure them.
- They are all manageable by standard medical care.
- All are lifestyle sensitive. In the early stages establishment of healthy habits can stop or even reverse the disease.

Those are bold statements. Surely we cannot be totally responsible for what we are! After all, we may have had a faulty set of genes dealt to us by nature, or we succumbed to some mutations (far more common than historically believed). While this may be true, it is also true that our genes are not carved into stone and how they are regulated and expressed is sculpted by repeated experience and environmental factors. Let's learn a bit about the genetics of AD, as a prime example, and we will discuss the impact of experience, the science of epigenetics, later in the final chapter.

Genetics

The brain's reaction to many types of injury is governed by genetic programs. A set of genes (ApoE genes, as noted earlier in Chapter 6) governs the formation, deposition and clearance of a substance called apolipoprotein. People whose brains have the inherent tendency to deposit much amyloid, a complex of apolipoprotein, or who do not have the mechanism to clear it out, accumulate these deposits in response to mechanical, vascular, inflammatory and other injuries. The result is a toxic reaction that kills brain cells and leads to more deposits and characteristic tangles of nerve fibres (neurofibrillary degeneration). The end result of this process is AD which is a slow killer disease, destroying its victim in about ten years. So we have two overlapping processes: amyloid deposition which is a genetic tendency to react a specific way to injury, and the injury itself, usually inadequate blood supply, inflammation or head injury.

Many risk factors for vascular disease also increase the risk of Alzheimer's disease. Brain damage from cell breakdown and neurofibrillary degeneration is complimentary with vascular damage.

Genetic and other tests for AD

ApoE genotype testing does not provide sufficient accuracy to be used alone as a diagnostic tool, but when used in combination with clinical history and clinical diagnostic criteria, it improves the accuracy of diagnosis. The gene expression is regulated by a variety of factors. We are just beginning to recognize and elucidate how these affect individual health. However, in the familial, early onset, strongly genetically controlled form of AD, the offspring have a much higher risk of getting the disease. In those cases genetic testing may be useful. Carriers, known or suspected, of the AD genotype need to know as they will need followup. Lifestyle changes may be necessary – avoidance of contact sports, lower stress, regular exercise, best nutrition, etc. Advanced care, family and estate planning may become relevant. Management of risk factors, as outlined in the rest of this chapter will be important.

Much research is being done to establish more definitive diagnosis for AD. It is just a matter of time before we will have a blood test to simplify this process.[1] Until then, the clinical exam will remain the most accurate way to establish diagnosis. Spinal tap to examine the cerebrospinal fluid and brain imaging

(SPECT, CT, MRI) remain backup diagnostic methods but they are not accurate enough by themselves as they do not show early changes.

Accurate, early diagnosis of degenerative brain disease raises ethical issues. What is the point of making an early diagnosis of a disease if we have no treatment? There are some arguments in favor of early diagnosis,[2] as well as arguments for a less rigorous diagnostic attack[3] – more about this later in this chapter.

Diagnosis of cerebrovascular brain disease

Cerebrovascular disease (CVD), causing poor circulation in the brain and brain cell death, is the second largest cause of major neurocognitive impairment. The good news is that CVD, unlike AD, is easier do diagnose and it is more modifiable. Major risk can be identified relatively early in the progress of the disease by the use of blood tests and brain imaging. Early intervention, as for cardiovascular disease, reaps benefits.[4] Although genetics does play a role, CVD tends to have multifactorial causation and risk factors are more easily addressed.[5]

Definitions and descriptions of the risk factors for early onset neurocognitive impaiment

I will now acquaint you with some of the independent risk factors for brain degeneration leading to neurocognitive impairment – the *Dirty Dozen*.

The term *independent risk factor* pertains to the fact that these risk factors act separately from each other, at least statistically speaking, and each one will independently and additively exert its own deleterious effect to increase further disease risk. In real life, of course, these are all intricately interrelated – not separate 'boxes' but more like windows onto the whole picture.

The term 'dirty' applies largely because, at least in the early stages, they develop insidiously, quietly without symptoms. They are 'silent killers'.

Alcoholism and substance abuse

Alcoholism and substance abuse syndromes are associated with brain damage. Moderate alcohol consumption may offer some protection against cerebrovascular and cardiovascular disease in some Caucasian populations because they have the enzymes to metabolize it and they may reap minor benefits from the antioxidant, relaxant and vasodilating properties of some alcoholic drinks, notably, red wine.[6] Overall, on a population basis, the harm appears to outweigh the benefit because alcohol is an addicting nerve toxin. As for other drugs of abuse, including marijuana,[7] they are toxic and not merely mind altering but brain altering – for the worse, not better. There is so much written about this topic that I will leave it at that. Anything that is a neurotoxin is bad for the brain.

Nicotine, in the amounts absorbed in cigarettes is not particularly toxic – it acts as a highly addictive stimulant. It is the carbon monoxide that deprives the brain cells of oxygen that is the culprit. On a more gradual chronic level, smoking simply destroys the lung tissues making it impossible for the lungs to do their primary job – to supply oxygen to the red blood cells which must have enough to give up to all other organs, including the oxygen-hungry brain. It also causes lung cancer.

The treatment for all substance abuse syndromes is withdrawal and abstinence. Given that substance abuse is a major destroyer of young brains, it is astounding that our lawmakers and health professionals do so little about them. It is almost as if they were on the side of the criminals who are pushing the drugs.

This sounds like an overstatement, but if we want to change the situation we should make it very difficult for people to obtain harmful drugs and at the same time make it very easy for them to get treatment if they become hooked. Today, just the opposite prevails. Drugs of abuse are far too easy to obtain and proper treatment is extremely difficult to find and to afford. Criminalizing such drugs is not the answer – education is! Effective treatment remains a major challenge.

Anxiety, stress and depressive disorders

Chronic anxiety disorders and depressive illness are the most common cause for a visit to the GP's office. As we have learned in Chapter 12, chronic stress associated with anxiety shrinks the hippocampus causing memory and emotional dysfunction. While that is not irreversible, it can set the stage for neurocognitive decline. Anxiety disorder is co-morbid with other mental disorders.[8] If untreated, it tends to progress to depressive illness.[9] The new way of conceptualizing depressive disorders is to think of the syndrome as a departure point for making more accurate diagnoses of underlying causes. Studies show that in some types of depressive illness the midbrain has compromised blood supply (ischemia).[9] This is a double whammy. Stress and the associated higher blood viscosity and ischemia feed on each other to create a vicious cycle of vascular insufficiency causing silent mini-strokes. In many cases of depression, the symptoms respond well to antidepressant treatments and do not progress to permanent neurocognitive decline. In almost all cases medical assessment is indicated to rule out treatable vascular factors and this is more important than mere symptomatic treatment with antidepressants.[10] Depression is no longer thought of as a single disease but a symptom complex that requires investigation and a differential diagnosis of causal factors. Depression is a biological state not simply a psychological one. There are specific SPECT and MRI findings and these change with treatment.

Some damage does appear to occur with lack of adequate treatment and repeated episodes of depressive illness. That leaves the victim with an increased sensitivity to stress and a higher chance of getting depression again. In any case, depression, particularly the late onset (after age 50) type, is a harbinger of probable brain disease and the development of major neurocognitive impairment.[11] It should trigger great concern, a thorough diagnostic workup and definitive treatment with regular follow-up.

Blood pressure disorders hypertension and chronic hypotension

There is really just one thing to remember about blood pressure: it has to be just right. It is easy to understand how blood pressure works if you think about a blood vessel as a hose. Imagine that you are watering a plant in your garden; if there is not enough pressure the water will just trickle out and you will not be able to water the plant properly. If the pressure is too high, on the other hand, it may blast the plant and damage it. If the pressure is high enough, for long enough, it may also damage the hose creating bulges or it may even burst. Healthy blood vessels are elastic and resilient, absorbing much of the shock waves accompanying the pump action of the heart. (More about the blood vessels later.) Both chronic high blood pressure (hypertension) and low blood pressure (hypotension) are associated with cognitive breakdown. Now don't get obsessed with your blood pressure because worry is likely to drive it up. Self-monitoring of blood pressure is a tricky activity because blood pressure changes according to various activities. Depending on what you are doing, working, walking, doing chores, watching television or sleeping, your blood pressure may be different under each of these conditions. Emotional states like fear and anger will also drive it up. It is also likely to be high when your doctor measures it because most people are a bit anxious when they are being examined.

It is important, however, to keep blood pressure in the optimal range. Your doctor knows how to do this but to do it properly you have to make a series of visits, not just one a year. Unfortunately, high blood pressure is very common and it is a bigger risk factor than hypotension because it is asymptomatic, hence the traditional name 'silent killer'. Hypotension, on the other hand, is likely to produce symptoms like lightheadedness, particularly upon standing up. As a result, it is seldom tolerated for long and remedies are sought. Hypertension affects approximately 50 million people in the United States and about a tenth of that in Canada.[12] Overall about a billion individuals are affected worldwide.[13]

Hypertension was, until a decade ago, the most common illness seen in a family physician's office but this was replaced by diabetes and more recently anxiety disorders and depressive illness have become number one. Hypertension is a risk factor for stroke, ischemic white matter lesions, silent infarcts, general atherosclerosis, myocardial infarction and cardiovascular morbidity and mortality.[14] The risk proportionately increases with increasing blood pressure but a high percentage of these vascular events also occur in those with normal blood pressure or mild hypertension. Several studies have reported that high blood pressure precedes AD by decades, but blood pressure decreases some years before the onset of major neurocognitive decline.[15] Blood pressure may actually be lower in individuals with AD than in those not afflicted.[16] High blood pressure has also been related to the neuropathological manifestations of AD. Hypertension often clusters with other vascular risk factors, including diabetes, obesity and hypercholesterolemia. Those risk factors have also been related to AD but more so to CVD. The exact mechanism behind these associations is not clear. Hypertension may cause CVD that may increase the likelihood that individuals with AD will express major neurocognitive impairment earlier in the syndrome. Hypertension may also accelerate the AD process – the neurofibrillary changes. Subclinical AD may lead to increased blood pressure – around and around it goes.

Hypertension is a common disorder and often left untreated. Even if hypertension only resulted in a moderately increased risk of AD or CVD, better treatment of hypertension would yield an immense reduction on the total number of individuals with major neurocognitive impairment.[17]

There are a number of authoritative studies which leave little doubt about the devastating effects of uncontrolled hypertension.[18] The Nun Study, quoted earlier, shed some light on the relationship between aging, hypertension and Alzheimer's disease.[19] Not all the women who met the pathologic criteria for AD at autopsy had suffered from clinical neurocognitive impairment during later life. The primary determining factor was presence of CVD. Those who had just one or two infarcts were over twenty times more likely to have suffered from major neurocognitive decline than those with none.[20] As noted, tiny silent infarcts throughout the brain are associated with hypertension. Each 1.0 mmHg increase in blood pressure over time increases the risk of poor cognitive function in later life by approximately 1%. Although hypertension is a well-known risk factor for strokes, it is also an independent risk factor for neurocognitive impairment. There are many other, even larger, studies that support the same conclusion: that hypertension is a major treatable risk factor in the development of neurocognitive impairment.[21] Almost everyone gets some of these tiny infarctions; therefore, the earlier prevention practices are adhered to, the better. The aim is to collect as few infarcts as possible.

That brings me to tell the good news. I like this part; I can almost feel my stress hormones and blood pressure going back to healthier levels. The good news is that hypertension is easily diagnosed and in fact it can be prevented. It is certainly manageable even after it has set in. There are many medications that have efficacy in stabilizing blood pressure in the optimal ranges. Recently, I had the honour of hearing an eminent neurologist colleague, Dr. Sandra Black, Head of Neurology at Sunnybrook and Women's Health Sciences Centre in Toronto, deliver a talk about modifying the progress of AD. She had explained on previous occasions that these tiny infarcts in the frontal lobes can occur when blood supply is disrupted as a result of hardening of the tiny arteries that feed this portion of brain tissue; the damage mucks up the connecting pathways that allow for what's known as executive decision making – the ability to plan and solve problems. These lesions "slowly strangle the brain", said Dr. Black, "But the good news is that damage can likely be prevented by taking the same steps as those recommended for warding off heart attack and stroke. Those include keeping blood pressure and cholesterol at healthy levels, not smoking, exercising regularly and eating correctly. It's very much related to risk factors that are controllable".[22] With all respect to Dr. Black, to this I would add, "stay well hydrated – drink plenty of water". You will see why when I explain blood clotting later in this chapter.

B-vitamin malabsorption and chronic deficiency

The discovery of vitamins has a very interesting history, particularly because the B vitamins are the most important for the brain. Dr. Casimir Funk, a Russian chemist, first coined the word *vitamin* in 1913. Early researchers believed that these compounds, which they called the 'vital amines', hence *vitamins*, found in some foods, were far more important than the caloric value of the food. They observed that cereals and legumes contained these vitamins. Polished rice, on the other hand, contained none of the vitamins because they remained in the polishings that were not eaten. Those who lived on a diet of polished rice developed a brain syndrome called *beriberi*. That was a type of syndrome with major neurocognitive impairment, characterized by memory impairment and the inflammation of various nerves leading to aches, pains and numbness. Today we don't see fully developed beriberi but the vitamins, actually the B-vitamins, remain very important because as we get older a certain amount of gastric atrophy takes place and the absorption of the B vitamins greatly diminishes. For example, it is possible to have a balanced diet with sufficient calories and still develop cyanocobalmin (vitamin B_{12}) deficiency. Folic acid (vitamin B_9) deficiency, associated with depressive mood changes, tends to occur in those who cannot consume enough fresh vegetables. Although folic acid is plentiful in vegetables and cereals, most people across North America have found it challenging to absorb enough; therefore, the government, both in Canada and the USA, have added it to cereals. However, despite the government boost, not all people can eat cereals and not all people absorb vitamins equally well. For that reason doctors should screen elderly, particularly those with memory problems for vitamin B_{12} and folic acid deficiency. Giving folic acid alone may be dangerous because it can mask vitamin B_{12} deficiency until it is too late to reverse its neurological damage.

That is why vitamin B$_{12}$ is always prescribed with folic acid. Often I will advise a vitamin B supplement even if no abnormality shows up because these vitamins are inexpensive, relatively safe and can prevent the rather common but serious deficiency syndrome. To make things a little more complicated, a substance called homocysteine, which we can measure in the blood, appears to be a marker for B vitamin deficiencies and/or faulty vitamin B metabolism. We do not know whether homocysteine causes damage in itself or if it is merely a marker for the phenomenon. However, we do know that chronically depressed vitamin B levels and elevated homocysteine levels are associated with vascular disease (atherosclerosis) in the brain, heart and other organs. I will go into a little more detail about the development of atherosclerosis later in this chapter when we get to the part on vascular diseases affecting the brain.

Coagulopathies: impairments in the tendency for the blood to clot

There are a number of common conditions that increase the blood's tendency to coagulate and clot. In the brain, heart and other organs, that leads to a small area of cell death – i.e. an infarction. To set the stage, poor circulation causes pooling of blood. This is called stasis and stasis leads to ischemia. This can take place in any part of the body but when ischemia occurs in the brain it may lead to small infarctions that cause the brain tissue to silently die. Conditions that contribute to stasis include lack of physical activity, dehydration, chronic stress (which increases the blood's viscosity) and too much fat and glucose in the blood. Estrogen, by dilating arterioles, can also have this effect if it contributes to stasis. The best way to improve blood circulation is to stay well hydrated (8 cups of water per day for a healthy active adult) and avoid a sedentary lifestyle. Stay fit and keep moving!

Some types of heart disease are also risk factors for brain disease – in particular, atrial fibrillation. This is a relatively common condition in which one of the pumping chambers, the atrium, of the heart becomes erratic and inefficient. The result is an increased tendency for blood clot formation in this chamber. This may be thrown off to arteries carrying blood to the brain. These little traveling blood clots will lodge in small arterioles causing blockage (embolism) and a shutting off of the blood supply. This is very similar to a coronary embolism, or heart attack, where it is the coronary artery that becomes blocked leading to the death of surrounding heart muscle – a heart attack. For the above reasons, irregular heartbeat and atrial fibrillation must be diagnosed and treated to minimize risk of 'brain attack' (stroke).

The opposite of too much coagulation is too little and that causes bleeding. Any medications or combinations of medications and herbal remedies that interfere with blood clotting may contribute to this hemorrhagic process. A number of substances that can be bought over the counter will interfere with the blood's ability to clot. These include aspirin, ginseng, gingko, ginger, willow and garlic extracts, to name a few. Their effects are additive. Doctors often prescribed low-dose aspirin to reduce clotting tendency in people with vascular disease. When the above substances are taken in isolation they are not likely to cause difficulties but there are always some people who think that if a little bit of something is good then more is better and these people may indulge in a number of these substances simultaneously, which may cause them to suffer a brain hemorrhage which is, of course, a stroke. *Caveat emptor!*

Diabetes type II - insulin resistance, high blood sugar, abdominal obesity and the metabolic syndrome (over-nutrition syndrome, Syndrome X)

Obesity is associated with chronic inflammation at the cellular level and this is a hazard for blood vessels as well as pancreatic islet cells where insulin is produced. Not surprisingly, obesity is another risk factor for diabetes and neurocognitive impairment whether Alzheimer's or vascular type. Unfortunately the risk increase is substantial – about double (100%) by midlife.[23] Although not all diabetics are obese, obesity is such a high risk factor that it belongs here as a kick-off point for the next part on diabetes. Diabetes is another risk factor, not only for heart disease (two- to fourfold) and peripheral vascular disease, but also for CVD and the resultant neurocognitive impairment.[24] Chronic diabetes has many nasty effects – among the most insidious is chronic vasculitis, or inflammation of blood vessels. As we described above, when this happens in the brain it sets the stage for mini strokes. Tiny areas of brain tissue damage will be followed by degenerative changes as described for AD.

Diabetes is a common but complex disorder. We covered it in some detail in Chapter 13, under the heading 'Insulin'. It is important to learn ways to beat diabetes – catch it early to stop it in its tracks:

- Test blood sugar
- Lower blood fats (cholesterol) if elevated
- Avoid obesity
- Stay physically active – exercise
- Monitor blood pressure to avoid hypertension
- Annual dilated eye exam to check blood vessels of eye (retina)
- Check feet for poor circulation, wounds and poor healing
- Get vaccinated every year. Diabetics are more prone to influenza and pneumonia
- Take a baby aspirin daily to reduce platelet clumping (clotting tendency)

- Diagnose and treat depression – it is linked with heart disease and diabetes
- When you see your doctor, take notes and ask questions
- Demand the best care – you are the captain of your health care team

Hormonal and endocrine disorders

Certain endocrine disorders such as thyroid disorder – the second most common endocrine disorder after diabetes, can cause a neurocognitive impairment which could be mistaken for a degenerative brain disease. However, hypothyroidism is reversible, most easily in the early stages; therefore, correct diagnosis and treatment are very important. There is some suggestion however that untreated thyroid disorder in itself may cause or predispose patients to degenerative brain disease. For more details you may refer back to the section on thyroid hormone in Chapter 13.

An interesting and increasingly common condition called 'metabolic syndrome' deserves some attention – again. This is also known as Syndrome X, Insulin Resistance Syndrome or Reaven's Syndrome, dicussed under the heading 'insulin' also in Chapter 13. Metabolic syndrome is associated with age, low physical activity, smoking, hypertension, high blood sugar levels, high cholesterol and central abdominal obesity (high waist-hip ratio – see Nutrition, Chapter 8). The American Diabetes Association and the European Diabetes Association are in agreement over the latter definition.[25] It is a very complex but common endocrine disorder. I think of it as pre-diabetes and I counsel my patients to take it seriously even if they have no overt symptoms or discomfort. It is associated with a number of nasty outcomes and if untreated will lead to coronary heart disease, hypertension, adult diabetes, obesity and stroke – otherwise known as CHAOS (See Chapter 8 – Nutrition, and Chapter 13 – Hormones). It is also associated with chronic inflammation of the tiny blood vessels or vasculitis which explains the coronary heart disease (CHD). There is also accelerated cognitive decline probably as a result of similar processes within the brain – cerebrovascular disease (CVD). Changing diet, activity level and other aspects of lifestyle can modify this condition.

Inflammatory disease of connective tissues

Every tissue in the body reacts to injury by an inflammatory response and the brain is no exception. It does not matter whether injury is mechanical as from TBI, from vascular disease, chemical insults or even chronic alcoholism. When I first started giving the Healthy Brain seminars in 2000, and was introducing the Healthy Brain method to a group of family physicians, it was my honour to have Dr. Patrick McGeer, from the Kinsmen Laboratory of Neurological

Research, University of British Columbia, as a keynote speaker for one of our symposia. The reason I felt honoured was not only because Dr. McGeer, having published over 500 scientific papers, was a pioneer in brain chemistry research in several areas of neuropsychiatric illness, but because he was well known for having discovered the link between a lower incidence of AD and the use of anti-inflammatory medications. He reported that people who suffered from painful arthritis and who had to consume large quantities of anti-inflammatory medications, had a significantly lower incidence of AD. For a more detailed discussion on this topic, I refer you to a chapter by Dr. Patrick McGeer and his favourite co-researcher, his wife. Dr. Edith McGeer, recently published in a book celebrating the 100th anniversary of AD research since the initial report of Dr. Alois Alzheimer on November 4, 1906 from Tubingen, Germany.[26] We owe our present knowledge that chronic inflammation is associated with a broad spectrum of neurodegenerative diseases of aging that includes disorders like AD, Parkinson's disease (PD), amyotrophic lateral sclerosis, multiple sclerosis (MS) and others, to their pioneering works.

The same applies to age-related macular degeneration of the retina – the leading cause of blindness. Although the retina is not considered to be anatomically part of the brain, this is an arbitrary demarcation. Physiologically, the retina is very closely related to brain cells because it is actually an outgrowth of brain tissue with specialized but similar nerve cells.

The inflammatory process, like the stress response (Chapter 12), has a double edge. In acute situations, or at low levels, it deals with injury and promotes healing; it helps to overcome minor damages. When chronically sustained at high levels, it seriously damages surrounding host tissue. The affected part of the body destroys itself. Epidemiological evidence overwhelmingly demonstrates that inflammation, on the long run, becomes very harmful in any organ system. So far, advantage has not been taken of opportunities indicated by these epidemiological studies, i.e. to use anti-inflammatory drugs to treat neurodegenerative conditions which are associated with inflammation. Locally produced self-destructive molecules are called *membrane attack complex* and comprise specific proteins and oxygen free radicals. Stimulation is provided by a variety of pro-inflammatory hormones called *cytokines*. Agents that reduce the intensity of inflammation should have broad spectrum application against degenerative diseases, after injury or as a result of pro-inflammatory risk factors accumulated during the aging process. Based on that evidence, classical non-steroidal anti-inflammatory medications (NSAIDs) are the most logical choice. Dosage must be sufficient to combat the inflammation. Analysis of levels of inflammatory mediators indicates that the intensity of inflammation is considerably higher in AD (hippocampus) and in PD (substantia nigra) than in osteoarthritic joints but unlike the joints, the brain feels no pain. Thus, full therapeutic doses of NSAIDs, or combinations of anti-inflammatory agents may be needed to achieve the suggested neurological benefits.

Unfortunately there are two problems. First, at this time, we do not have accurate, reliable blood tests for inflammation of the brain, but there are various ways to check out inflammatory activity in the body. I advise that chronic inflammatory disease markers should be ascertained in middle aged folk and if inflammatory disease starts to rear its ugly head it should be stopped early because we know that elevation of pro-inflammatory chemicals will also affect the brain – without even any pain! Secondly, NSAIDs have side effects. They can cause gastric atrophy and bleeding, and even acute hemorrhage. Prescribing these medications requires expertise and medical supervision for side effects. Ask your doctor about these; he will know what you mean. Then insist on a full diagnostic workup. Often a rheumatologist is the next logical person to consult. There will be breakthroughs in anti-inflamitory treatments for neurodegenerative disease but we are not there yet.

Lipid disorders – fat in the blood

Fat on the body, fat inside the abdomen around the viscera and fat in the blood are not all the same and are not always present simultaneously. The brain itself is almost all cholesterol. This is so because nerve cell membranes are fatty substances and the surroundings of the nerve fibers, the myelin, are also related to cholesterol. So, fat can't be all bad. Particularly, the developing brains of children actually need a lot of fat. Did you know that human breast milk has a very high fat content? Our blood also has many types of fat circulating in it. What counts is the ratio of 'good cholesterol' and 'bad cholesterol' – good fat or bad fat in the blood. Not only should the good fats outweigh the bad fats but they should not attach themselves to the blood vessel walls. That is the beginning of atherosclerosis. Again, I note that much of the research is focused on cardiac rather than brain vascular disease. Hippocrates was right when he said 2,500 hears ago "What is good for the heart is generally good for the brain". Can we say the converse, "what is bad for the heart is generally bad for the brain?" That position is well supported by recent research. Thus, high cholesterol increases risk of AD and CVD.[27] Smoking, hypertension, high cholesterol and diabetes at midlife were each independently associated with a high increase in risk of neurocognitive impairment[28] and the combination of high cholesterol and high blood pressure were particularly hazardous because they also increase the risk of AD.[29]

Periodontal disease

It is interesting to note that dentists have been telling us for years that periodontal disease (chronic gum infection, gingivitis) is associated with small blood vessel inflammation which leads to coronary heart disease. Is this why cowboys, farmers and horse traders always check a horse's mouth before buying it? Chronic inflammatory disease, either caused by or mirrored by chronic gum infections, turns out to be a good indicator of poor general health. It appears that when infection attacks the gums, the germs and toxins produce a vascular inflammatory response in the blood vessels supplying the heart itself. When these small arterioles in the heart muscle are inflamed, they develop clots and become occluded, which leads to a heart attack. That process does not stop at the heart. Studies now suggest that a similar process may affect brain tissue.[30] It appears that chronic gingivitis is also associated with CVD and stroke. Think about it; it is not surprising that this same process is not restricting itself to the heart, but repeats itself in other tissues including the brain.

Take oral hygiene very seriously – get you teeth and gums cleaned by your dentist, floss after meals and brush your teeth in the morning and night. Oral bacteria love glucose and other sugars found in our diet. Xylitol, an unusual (5-carbon) but safe natural sugar from plants is effective in reducing gum infection and tooth decay.[31] It is available in food stores. Even if you start using xylitol, don't kiss your dentist good-bye as you should continue with checkups and dental cleaning as needed.

One more word about your dentist – he or she can save your life! Your dentist may be the most important doctor you see for a checkup because dentists routinely spot signs of cancer, diabetes, heart disease, Crohn's disease and less common but serious rare skin and autoimmune diseases. Smile!

Poison exposure and ingestion

There are toxins, in particular pesticides, used mostly in rural areas that are associated with cumulative brain damage. Mercury is highly toxic to the brain and even small amounts will damage brain tissues. The fetus in the uterus and children are highly susceptible. Lead poisoning tends to be cumulative from gradual repeated exposure and is also a risk factor for neurocognitive impairment and brain damage.

Lead poisoning can occur from toys. In 2004, 150 million toys, bracelets, rings and necklaces have been recalled by the Consumer Product Safety Commission because they contained dangerous amounts of lead. That was the largest recall ever issued in the US. There are 700,000 vending machines that dispense toy jewellery in the USA. Again, in 2007 history repeated itself with major recalls of toys manufactured in China.[32]

Physicians do not advocate routine screening for lead and other toxin levels because they are found in almost everyone in very small amounts. Epidemiologists have tried to define 'tolerable levels' but recent research supports the conclusion that there is no tolerable level – the lower the better and the ideal would be zero.

This is an inconvenient and unpleasant truth. However, high risk groups do need to be tested. For example, recent immigrants, particularly children from China may have unacceptably high lead levels and may require medical treatment.[33]

Some occupational and environmental hazards are also relevant. People on well water may accumulate arsenic or other toxins including heavy metals. People who eat a large amount of fish in contaminated regions should be tested for mercury. Workers, like welders, exposed to metal powders, particularly manganese, may develop neurological disease. The list goes on.

To add insult to injury, there are many medications prescribed by us, as doctors, with toxic side effects which may, at least temporarily, interfere with neurocognitive function. Sedative hypnotic medications and antihistamines used for their sedating effects will cause cognitive impairment, especially in elderly people even if they do not have neurocognitive impairment to start with. Fortunately that is usually reversible. A particularly deleterious type of side effect called *anticholinergia* is very common because there is a large and widely used list of drugs that have anticholinergic effects and these effects are additive. Anticholinergic compounds, the most powerful one being atropine, oppose the effects of the neurotransmitter acetylcholine in the brain. Acetylcholine plays a major role in concentration, thinking, perception and even consciousness. Opposing this vital neurotransmitter, which tends to diminish with age anyway, anticholinergics can cause severe cognitive impairment, 'brain fog', even delirium and death. Their effects are *additive*, which means that if you are on more than one, the effects are added together. A partial list of drugs with anticholinergic side effects follows:

Alprazolam	Amantadine	Amitriptyline	Ampicillin
Atropine	Azathioprine	Captopril	Cefamandole
Cefoxitin	Chlorazepate	Chlordiazepoxide	Chlorthalidone
Cimetidine	Clindamycin	Codeine	Corticosterone
Cycloserine	Cyclosporin	Desipramine	Dexamethasone
Diazepam	Digoxin	Diltiazem	Diphenhydramine
Dipridamole	Dyazide	Flunitrazepam	Flurazepam
Furosemide	Gentamycin	Hydralazine	Hydrochlorathiazide
Hydrocortisone	Hydroxyzine	Imipramine	Isosorbide
Keflin	Lanoxin	Methyldopa	Nifedipine
Oxazepam	Oxybutynin	Oxycodone	Pancuronium bromide
Phenetzine	Phenobarbitol	Piperacillin	Prednisolone
Ratinidine	Theophylline	Thioridazine	Ticrocillin
Tolterodine	Valproic acid	Vancomycin	Warfarin

When you get one or more prescriptions, ask about side effects affecting the brain. The term *side effect* is usually just that, a minor, sometimes even unpleasant unintended effect of a drug prescribed for its main, important intended beneficial effect. Side effects are not always present, may be due to suggestion or may be tolerated in favour of the main effect. With the brain however, a side effect can mean more than an inconvenience.

Vascular disease and atherosclerosis

Vascular disease itself, especially in tiny arteries, will lead to coagulation because inflammation of the lining of these small blood vessels exudes chemicals that will clot the blood. Vasculitis can have many correlates but it is bad enough all by itself. When it is accompanied by elevated lipids (bad cholesterol), homocysteine, glucose, elevated clotting factors from chronic stress (remember cortisol) and hypertension, then it really takes on a galloping form. Chronic damages to the inside lining of the arteries cause scarring. Calcium, fats, clotting factors and other goop are deposited, build up and may completely block the vessel. During this process the natural elasticity of the blood vessel is lost and that is why atherosclerosis is called 'hardening of the arteries'. When these events take place in the brain it is referred to as cerebral ischemic disease or cerebrovascular disease (CVD). CVD is characterized by transient ischemic attacks (TIAs) or frank stroke.

Migraine conditions, characterized by intense pain when blood vessels in the brain contract and then dilate, are a risk factor for poor brain circulation, TIAs and strokes. About 85% of strokes are caused by poor circulation or blood clotting in the brain, particularly in the small blood vessels. Most of these strokes are silent and because they are very small they do not cause symptoms but cumulatively they do cause damage and lead to what is called major neurocognitive impairment of CVD. Note that neurocognitive impairment from CVD refers to a process, not a region of the brain. Thus, one can have CVD affecting mostly the frontal lobes, temporal lobes, hippocampal areas or any combination of regions of the brain. The other 15% of strokes are hemorrhagic, meaning that they are caused by bleeding into the brain tissue. That may be caused by high blood pressure bursting some small arteries or breaking an aneurysm which is a weakened and bulging area of the blood vessel. Too much anticoagulation may be a factor.

Recognition of CVD and treatment remain extremely important. Signs may be absent or present as migraines, dizziness, transient blackouts or loss of function in speech, confusion or some temporary disturbance in perception.

Mild cognitive imparement (MCI), which until recently was called 'cognitive impairment, no dementia' (CIND), is a syndrome often associated with silent underlying diseases like diabetes and vascular disease that must be recognized and treated. MCI syndrome is common, particularly in the over 65 year old age group and nursing home residents.[34] It is usually of vascular origin and it can improve dramatically with treatment. The treatment of risk factor diseases is more effective and more important than treatment with cognition enhancing drugs. That is a very important diagnostic distinction because AD is not like vascular dementia in this respect. Aggressive treatment is recommended for stroke prevention because even in the absence of a big stroke, tiny silent mini-strokes can take place. The cumulative effect, over months or years, may be major neurocognitive impairment. The distribution of ischemic strokes can be divided into large vessel thrombosis 25%; cardiac embolism 25%; and small vessel occlusion 25%. Approximately 19% are due to undetermined ischemic causes and the remaining 6% a result of other defined ischemic causes. I must emphasize that the most important established risk factors for small vessel disease are hypertension and diabetes. An inherited tendency, or family history of hypertension, diabetes, smoking and high levels of fat in the blood (hyperlipidemia) also increase the risk. Having a previous cerebrovascular accident (CVA), meaning a stroke or TIA, an MI or peripheral vascular disease greatly increase the risk of further attacks.

Clinical implications of atherosclerosis[35]

Initial Event	Increased risk of MI	Increased risk of CVA
Cerebrovascular accident (CVA)	2-3 x	9 x
Myocardial infarction (MI)	5-7 x	3-4 x
Peripheral artery Disease	4 x	2-3 x

NB: A twofold (2x) increase = 100% increase

The problem in trying to understand vascular disease is that the vasculature, a complex organ in its own right, is usually artificially cut up into three segments: the neurologist looks after the brain, the cardiologist looks after the heart and the vascular surgeon looks after the legs. However, damage to the blood vessel linings and plaque formation, rupture and clot formation may be happening everywhere, in every artery of the body. It is therefore not surprising that some cerebrovascular disease is a common finding in normal elderly individuals. The prevalence of internal carotid artery atherosclerosis in elderly individuals in the community is estimated to be between 23% and 34% overall.[36] CVD is also a common finding in patients with other manifestations of artery disease and these individuals are at increased risk of experiencing a vascular event, such as a clot in the leg, heart or brain (stroke). In fact, peripheral artery disease, say in the feet, legs or fingers, is also a strong indicator of a similar process in the brain. 10% of patients with symptomatic peripheral artery disease have internal carotid artery stenosis greater than 70%.[37] In these patients the annual rate of stroke or TIA rate is 3-5%. 10-17% of patients with aortic aneurysms larger than 25 mm in size have internal carotid narrowing greater than 50%. Their annual risk of stroke or TIA is also 3-5%. About 40% of patients undergoing coronary heart disease surgery are found to have internal carotid narrowing of at least 50%.[38]

Clearly it remains extremely important to keep our blood vessels healthy and open. Healthy tiny arteries and even larger vessels are able to contract and expand according to demand for blood flow. Healthy vessels do not have fatty and waxy materials embedded in their walls and of course, they are not narrowed or occluded. The development of atherosclerosis in the early stages appears to be related to the deterioration of the delicate lining inside the blood vessels.

Homocysteine is a naturally occurring sulfur containing amino acid and researchers believe that increased levels of this substance in the blood stream contribute to the breakdown of the cellular lining and connective tissue in blood vessels. In short, increased homocysteine is associated with evidence of vascular lining injury. It also appears to be associated with thickening of the arterioles and progressive arteriole stenosis. Homocysteine is generated by a complex chemical reaction that is governed by the B vitamins. As I noted earlier in this chapter, chronically low levels of B vitamins appear to contribute to elevated homocysteine levels.[39] Other factors associated with increased homocysteine levels include age, male gender, high dietary protein (methonine) intake, menopause, vitamin deficiency, renal disease, transplantation, cancer and hypothyroidism.[40] A number of medications such as corticosteroids, theophylline and some anticonvulsants, cyclosporine, methotrexate and smoking also increase homocysteine levels. Homocysteine levels tend to be lower during pregnancy, during the administration of B vitamins (folic acid, B6 and B12), with low albumin (a shortage of protein), in post-stroke states and severe psoriasis. Oral contraceptive hormone replacement and the drugs penicillamine and N-acetylcysteine tend to lower homocysteine levels. AD and vascular dementia are associated with elevated homocysteine levels.[41] Recent research has shown strong evidence that increased homocysteine levels, or low B vitamin status, even within the current 'normal' reference ranges, may increase the risk of developing neurocognitive impairment.[42] The association of high homocysteine levels with vascular disease has been thoroughly documented.[43] Several prospective studies have also shown that elevated levels of homocysteine increase the risk of cerebrovascular disease.[44]

We learned earlier that CVD has been considered the second most common cause of major neurocognitive impairment after AD. High homocysteine levels are however, associated with both CVD and AD.[45] Vascular disease may lower the threshold for overt cognitive impairment caused by mechanisms specific to

Alzheimer's type neurofibrillary degeneration. When patients complaining about cognitive disturbances were investigated using a number of diagnostic methods, the following was found: 33% of the patients with subjective memory complaints, 45% of patients with cognitive impairment of AD and 62% of the patients with cognitive impairment of CVD had high homocysteine levels.[46]

Several pathogenetic mechanisms linking homocysteine to major neurocognitive impairment have been proposed but none have been proven and it may be that they all act concurrently.[47] Results from planned or ongoing intervention studies will determine to what extent that may be true. Oxford University's Dr. David Smith recently commented in an *American Journal of Clinical Nutrition* editorial: "If lowering homocysteine could stop only 10% of those with mild cognitive impairment from developing AD, several hundred thousand persons worldwide would benefit every year".[48] It has been established that there is an inverse relationship between B_{12} and folate levels and homocysteine in the blood; the lower the B vitamins the higher the homocysteine.[49] Some studies have shown that lowering the homocysteine level improves cognitive performance[50] but some have not been able to demonstrate that.[51] We cannot draw major conclusions about the role of homocysteine but an optimal B vitamin status remains important.

Other nutrients, for example vitamin C (ascorbic acid), are also important for blood vessel integrity. Vitamin C plays an important role in the development and maintenance of connective tissue. It is believed that with suboptimal levels of vitamin C, the connective tissue of blood vessels deteriorates and the defective surfaces attract lipids and sugar clumping to produce atheroma formation, which is atherosclerosis.[52]

In summary, high blood fat (cholesterol), interacting with sugar and high homocysteine clumping in blood vessels with damaged connective tissue appear to be the process of atherosclerosis.

Prevention

All of the above twelve conditions which I nicknamed the 'Dirty Dozen' are risk factors which can be detected in the early stages. Treatment exists for each and all of them. It is much easier to reverse, or at least stop the progress of pathology when those conditions are caught early. What is most disconcerting is that damage can and does occur silently in the early stages of the above common diseases. If any of the above conditions are detected, causes should be identified and hopefully eliminated. If that does not fix the problem, an assertive treatment program should be instituted.

Section II: General treatment issues

There are a number of problems that have to be overcome before a person can find treatment for the *Dirty Dozen*. First of all, one has to find a sympathetic and collaborative physician who will screen for the disorders I have described. With a careful history and some diagnostic testing it is easy to identify each one before it reaches late stages. Following that initial screening, the physician will have to enter a process of diagnosis and quantification of the suspected illness. Ideally, then the physician and patient will embark upon a journey of collaborative work to stabilize or, in the early stages, even reverse some of these conditions. When this works, the risk for neurocognive decline is greatly reduced. It is not only the doctor who has to take responsibility for the collaborative process, but also the patient. We know that people do not like to be patronized and infantilized by a high handed paternalistic approach. Personal autonomy is important in our culture. We also know that treatment adherence improves when patients' concerns are taken into consideration and treatment plans are carefully negotiated.[1] That is why I say that your brain is your best health care system. Again, you are the owner of your brain.[2] Even after acknowledging the above, the plot seems to thicken.

Diagnosis and treatment – collaboration vs. defensive medicine

Immediately, new obstacles surface. First of all it is difficult to find a physician who has time to collaborate with patients. The fee structure is such that visits have to be short and to the point. Complex problems tend to be shunned – and some patients are only too happy to do so! Unfortunately, there is a shortage of generalists and family physicians with a broad understanding of illness. There have been serious efforts to promote collaboration, not only between patients and their physicians, but also between general practitioners and specialists in brain disorders. That was the 'shared care initiative', which may be operating under a variety of names but is alive and well in some regions.[3]

Over the last few decades, since Dr. Michael Balint proposed his viewpoint, there has been a shift, at least in the UK, toward *Balint Groups* as an aid in continuing medical education. Balint Groups are regular meetings of a group of physicians over an extended period of time. There is no teaching, no agenda – only open discussion. These physicians make a commitment to meet in groups of ten once a month for a minimum of ten sessions. The material worked on is from current cases. The purpose is to increase understanding of how the patient's emotional problems affect both the illness and the patients themselves. The leader preferably has psychoanalytic training. Balint Groups have spread since the 1950s and exist in most countries. Most would agree that there is a need and demand for a forum where physicians can gain empathic, intellectual support and increase their sophistication in understanding *patient-doctor-illness* interactions.[4]

Unfortunately, that is not all that is needed because in addition to physician-patient factors, there are political, legal and market forces at play.

There are also difficulties in diagnosis and treatment, not just because of the inherent diagnostic and technical problems related to the illness, but also because of patient factors and psychological factors pertaining to decision making. Doctors often have to make decisions quickly, but that is not their downfall. The greatest obstacle to making correct decisions is not lack of time but distortions and biases in the way information is presented, gathered and assimilated. Doctors do try to train themselves in how to overcome pitfalls in decision-making but patients, as consumers, have little preparation and understanding of how to present information, in the most useful way to promote their doctors' capabilities.

Under difficult times in the delivery of health care, the practice of *defensive medicine* rears its ugly head. Physicians may practice defensive medicine without guilt because the action is either unacknowledged or not considered unethical. Not having time to collaborate and problem solve, doctors have to fall back on the overuse of tests and procedures to regard themselves and to be regarded by others as being thorough and complete in their approach to diagnosis and treatment.

Particularly absurd, for physicians and patients alike, are the present day trends of patient expectations. Unable to tolerate uncertainty, many patients expect and pursue multiple treatment modalities for the same problem, arriving at treatments lacking methodology and which are not based on knowledge of clinical medicine and scientific evidence. That approach, it has been noted, adds significantly to the enormous expense of contemporary health care.[5] When unrecognized, the above 'scrambled eggs' approach yields medical outcomes that may be unaccountably changed for the worse. In the USA, 'multiple doctoring' and the pursuit of whatever is advocated in the health care marketplace obviously translates into big dollars – the huge health care industry. In countries where health care is 'covered' and the illusion is that it is free, the cost of inefficient health care is merely hidden only to declare itself as high taxes, political turmoil, long waitlists and generally inadequate health care delivery.[6]

I remarked in previous chapters, that many diseases associated with aging are on the rise because we are living longer. We accumulate risk factors as we go along. The brain remains the weakest link in organ repair and replacement. As a result, brain related diseases are reaching epidemic proportions. The main risk factor for major neurocognitive impairment is aging itself and that we cannot control. We tend to accumulate risk factors as we age, but don't lose heart; healthy aging is possible. Major neurocognitive impairment is a result of real disease and it is not part of healthy aging.

Medical care is in crisis

At the present time there is a crisis in medical care. It is beyond the scope of this book to explore the subtle causes but it is important to say that it is happening not only in the USA and Canada but also in other countries. For complex reasons patients are less cooperative with physicians than they used to be. What is the nature of this ambivalence? On one level, people yearn for the doctor (the original meaning of the Latin word *doctor* was *teacher*), the kind authoritative advisor, yet on another level we resent paternalism, control and interference with our autonomy. The doctor of 'the good old days' was an idealized, powerful figure who delivered babies, looked after kids, their mothers and fathers and the whole family. They were highly regarded, not likely to be exploited or abused and when they died, many patients went weeping to their funerals. The doctors of 'modern times' are in a very different position. They are not trained in the 'art of medicine' but operate more like scientists and technicians. Medical science has progressed so much over the last generation and there is so much more knowledge and information to manage, that the average physician is overwhelmed. Doctors today have learned the scientific method, think in probabilities, not facts, and like all scientists, try to remain skeptical and objective. They manage uncertainty.[7] Perhaps that is why there is an over-reliance on drugs.[7.1] Drug therapies appear to reduce unpredictability but may actually increase chaotic outcomes.

As a result, the effectiveness of physicians has declined. Patients now question what doctors are saying, and that may be a good thing. If communication breaks down, however, it isn't a good thing. There is an anti-intellectual climate and a proliferation of 'junk science' and that is supported by consumerism as well as by under-informed and power-hungry governments.[8] Again, *caveat emptor*!

We will look at the pressures on doctors and health care delivery from three perspectives: social factors, doctor factors and patient factors.

Social factors complicating health care delivery

People are more knowledgable than ever. However, through the media and the Internet we, as patients are overwhelmed by medical information and misinformation. Typically, we have more time to amass information on whatever ails us than the doctor! It is not unusual for me to see a patient who has done 'google research' and presents with all kinds of facts and factoids about their 'syndrome'. A well informed patient is very helpful but a gullible one, full of 'junk science' and superstition is a real chore.

The sheer complexity in biomedical knowledge has made a commonsense approach very difficult. The trend toward specialization has squeezed out the family physician and generalist as the most important member of the healthcare team. Lack of respect and appreciation has caused physicians to migrate to greener pastures or quit practice altogether. The atmosphere of mistrust, at least in the USA, leads to practicing defensive medicine which costs tens of billions of dollars per year.[9]

The above observations bring up the question whether we need more and more healthcare or do we need more and more prevention. In Canada, as in other countries, the total drug expenditure (public and private) as a percentage of total health expenditure has been steadily increasing from about 10% in 1985 to 15% in 2000, and so forth.[10] UK Health Minister, Brian Mawhinney reported in 1995 after a study of total drug use through a large home survey, that 150 tons of prescription drugs went to waste each year.[11] Do you think the situation has improved? Can drugs give us a long Health Span to match a long lifespan?

Social trends have promoted autonomy – overall that's good. People now have unlimited access to medical information and have great expectations. In the romanticised 'olden days' doctors had *patients* (from the Latin *patiens* meaning to suffer, endure). Now we have *clients* or health care *consumers*. The illusion is that we can just buy health, or at least health care and that we have a right to the magic of health care. However, health is not really a commodity and health care has always been limited. To get a better understanding, I suggest you see a play by George Bernard Shaw called the *Doctor's Dilemma*.

Doctor factors complicating health care delivery

Recently I saw a cartoon that showed a physician informing a patient that his specialty was diagnosis and now he would have to refer the patient to another doctor whose specialty was treatment. Although that was a joke, unfortunately this sort of thing does happen. Surveys show that that not only patients are disenchanted with medical practice – physicians are also. Physicians complain of high workload. Studies show that high workloads may impair a physician's judgment.[12] Under conditions of stress and poor sleep, physicians and hospital staff are prone to what are politely called 'slips, lapses and trips'.[13] Doctors feel frustrated if their patients are not getting well.[14] This is a big problem in the treatment of chronic illness and in the treatment of neurocognitive impairment which is often chronic. Most physicians complained that they do not have adequate time for discussion of problems with their patients. Family physicians are in a particularly stressful position because they have to manage uncertainty while faced with an unhappy and suffering patient as they remain powerless until the disease unfolds to a point where it becomes more specific, diagnosable and treatable.

Another source of physician dissatisfaction is the sheer volume of paperwork and record keeping. The amount of time a physician has to spend filling out forms, being on the computer (electronic records) or being 'on hold' on the telephone is a serious threat to the time available for patients – and patients, or shall I say consumers, understandably hate to be kept waiting or be faced with a doctor who only gives them 5-7 minutes, will only hear one complaint and makes eye contact only with his computer monitor. I may be exaggerating a bit – but it's to make a point. Many doctors do feel that mandatory electronic medical record (EMR) keeping is detrimental to the doctor-patient relationship.[15]

Patient factors complicating health care delivery

Other obstacles to treatment pertain to patient profiles. Non-adherence to treatment is a serious waste of time both for doctor and patient. It is also a waste of money and resources and may put the patient at further risk of complications. Patient factors predicting non-adherence to treatment include thefollowing:

Low self-esteem	Unstable lifestyle
Substance abuse	Younger age
Male gender	Negative family attitudes
Grandiosity, suspiciousness	Negative attitude to medication
Poor insight into difficulties	Lack of organization

Interestingly, when doctors and patients are surveyed, they generally have overlapping expectations of each other. They would like to partner and communicate and to have time to do that. Doctors would like to spend a little more time with their patients and vice versa. Patients generally expect a client-centered treatment approach. About 25% of patients expect a prescription; 60% of patients expect a physical examination.[16] Please note that in some cultures a 'prescription' is not merely a pill but may be a food, diet, massage, exercise or some form of behaviour. As patients, we are too accustomed to the magic of pills. Given that the doctor has only a few minutes, no wonder both patients and doctors are frustrated.

Medications and medication prescribing can also contribute to nonadherence to treatment. Polypharmacy or to use of multiple medications leads to complex, impossible to follow medication regimens, drug interactions and unexpected side effects. Cost of drugs may be a formidable obstacle.

We tend to believe that there is a pill for everything but, given that there are many ailments and illnesses, there must be a limit to how many pills one can take in a day. The very fact that we are presented with a huge choice of pills for every imaginable symptom leads to complex medication regimens that are almost impossible to follow. As a result the noncompliance, or non-adherence to medical treatment, costs the Canadian taxpayer an estimated $7-$9 billion.[17]

It is my recommendation that you, as the patient, reclaim ownership of your health care and in particular your brain health. You are not a consumer or a customer, but a stakeholder and collaborator in your treatment. You should be the leader. In partnering with your physician, treatment is hard work for both of you. It is important to define a goal. Do not remain a passive recipient of prescriptions and instructions and always make sure that you understand what you're supposed to do, or not do. Some of this may seem like common sense but it is certainly not common practice. Medical work remains complex and your brain remains your best healthcare system. Keep in mind that early diagnosis is important. Become proactive, set goals and measure your progress. For all that you will have to actively collaborate with your physician. It's not just *what* is prescribed but *how* it is prescribed and how you as the patient feel about the whole process that will determine the effectiveness of the doctor-patient collaboration.[18]

AAA treatment = Accurate diagnosis + Appropriate prescription + Assertive treatment

THE LEAST YOU NEED TO KNOW

- The Dirty Dozen predispose us to brain disease and are the risk factors for early onset neurocognitive impairment aka dementia.
- They can be easily diagnosed and treated – the earlier the better.
- It is easier to arrest or even reverse these disorders in their early stages.
- It is important to be proactive, not passive.
- Collaboration with your doctor or healthcare team leads to mutual trust and better treatment effectiveness and better outcome.
- The goal is to increase Health Span and quality of life not merely to increase lifespan.

START NOW!

- Get a checkup from your doctor and from your dentist! Do not ignore risk factors for brain disease.
- Be proactive – seek early diagnosis and insist on AAA treatment.
- Maintain your autonomy but collaborate with your doctor or health care team.
- Use your head: keep basic records and goals – share them with your doctor.
- Find ways to measure outcome – write them down.
- Monitor your progress.
- Know what medications you are taking and why – write it down.
- Know the fundamental effects and side effects of your medications.

www.HealthyBrain.org

EPILOGUE

...[it's] like what happens when we separate a jigsaw puzzle into its five hundred pieces: The over-all picture disappears. This is the state of modern medicine: It has lost the sense of the unity of man. Such is the price it has paid for its scientific progress. It has sacrificed art to science.

– Paul Tournier, MD.

The practice of medicine has changed – for the worse – over the last few decades. There are many reasons for that and doctors are just as much victims as agents of the change. The end result is that our physicians are overwhelmed with spurious scientific information, sheer numbers of patients, especially elderly patients with complex problems and have very little time (average seven minutes) per patient.

As with science in general, medicine has had to adapt to increasing complexity by increasing specialization. This is dangerous for us as patients or health care consumers. *Caveat emptor* was not a term applied to organized medicine – but now it is wise to beware. In fact we are no longer even consumers. We are 'triaged' – a process that has grown out of the battlefield, not civilian medicine. We are recipients of rationed health care and that has some pretty ugly implications and some pretty harsh practitioners.

The most insidious, and least recognized effect of specialization is that it promotes the concentration of knowledge into subject-specific 'guilds'. Medicine has become like a professional game or sport where a player's worth is assessed solely within a narrowly defined, often spectacular, event. When a discipline becomes a 'guild' it ceases to put value on communicating with other guilds and even less so with the public at large. This is all very natural, but highly undesirable for us as health care consumers – especially when we are sick. In its ultimate caricature form, your doctor is a person who knows more and more about less and less, does not talk to other professionals and has no time to listen and make clinical observations of you – the patient. He just knows what to do; like a 'one trick pony'. The guild-based nature of subspecialized medicine and the decline of the wise general practitioner led to a loss of public interest and trust. Benoit Mandelbrot, a profound scientific philosopher (the father of fractals), became very concerned about our fate when he made the discovery that modern science is now becoming too specialized, too fragmented; well written up in *The Economist Technology Quarterly,* December 6, 2004.

What has this to do with the brain? Plenty. One needs knowledge and good judgment on procuring and maintaining optimal health care. It is not a passive process that can be left up to another expert (guild doctor) or the government (triage officer). The consumer has to be well informed, assertive and appropriate in entering the health care field. Only a good brain can do that.

Medical science requires the approval and trust of society, which depends in turn on credible and articulate teachers who can skillfully engage the isolated ivory tower specialists. We intend to be such "interdisciplinary mavericks" (Mandelbrot's term) and we urge you, the reader, to develop a similar philosophy. Strive to break down the barriers between the 'guilds' and reconnect them with the wider, real world. Use advocates (friends, family, caregivers) and the legal system if you must. As for us tamer individuals, it behooves us to try to make sense of the growing variety of treatments and practices available and to make some effort to understand interactions that promote better brain health. Putting a high value on brain health starts before birth and in the care of the baby. It includes education. It remains highly relevant in work life and finally in coping with old age.

Today, doctors are expected to be like skeptical scientists who think in terms of probabilities and levels of uncertainty. This is great in pure research but very bad for the dynamics of physicianhood. Patients who come for help are not part of a trial or experiment; they do not wish to feel like 'guinea pigs'. They need guidance and reassurance based on definitive diagnosis and treatment plan. While physicians, especially family docs (and their patients) must often deal with uncertainty, they cannot only rely on science – that is just not appropriate. A competent experienced doctor has more in common with a successful artist or business man – they operate on pattern recognition, using hunches, not facts and factoids. They are not linear but lateral thinkers and synthesizers – they come up with solutions. Even our laboratory tests, which try to make scientific measurement of body functions as factual as possible, suffer from an error rate of about 15%. That does not seem too serious but it does mean that if your doctor orders a battery of seven or eight tests, one will probably show an erroneous result. To be truly rigorously scientific, the scientist-physician would have to repeat the tests over and over, ad infinitum, seeking certainty – and I'm afraid some of this is really going on.

Another bugaboo today is 'evidence based medicine'. It is reaching methodological fetishistic proportions; a pseudoreligion amongst medical scientists and what is worse, amongst some doctors who are not scientists. There are many treatments and approaches in the practice of medicine that cannot be measured and quantified – there may never be 'evidence'. The randomized controlled trial (RCT), the gold standard of research, opens windows into some processes and provides great insights but cannot investigate 'real world' situations. The RCT cult may set artificial limits because some clinicians may take 'absence of proof' to be 'proof of absence' of efficacy of medical treatments. That 'evidence based' approach can lead to negligence because, instead of applying what is empirically known, it encourages the physician to stick to a credo of defining higher and higher levels of certainty, instead of getting on with applying what is accepted as generally workable.

When a doctor and patient approach each other they should be reminded that, "...to be conscious that you are ignorant is a great step to knowledge", in Benjamin Disraeli's words (1804-1881). "Our profession, after all, deals partly with guess work; we do not deal in absolutes", said Dr. Paul Beeson.[1] To this I would add that our profession deals mostly with disorders that are multifactorial in origins; therefore, the seeking of higher and higher levels of certainty and the insistence on a 'silver bullet' can become a futile exercise which may deprive millions of people from getting basic health care – we need effectiveness, not just efficacy. We must recognize that the mind and brain (in fact the whole body) is a *chaotic system*, a non-linear dynamic system, interacting with the environment.[2] Inner rhythms, functional units and processes have fuzzy boundaries. *Fuzzy logic* (from chaos theory), as a many-valued logic, deals with reasoning in approximates rather than with exact rules.[3] Interactionism, comprehensive and complex management plans, like The Healthy Brain Program (or Healthy Heart Program), are more workable than simplistic prescriptions. The brain is complex and loves complexity; it sucks order out of chaos. Whenever possible, elements of The Healthy Brain Program should certainly be based on high levels of scientific evidence. We cannot ignore scientific thought and new developments for they will always have much to contribute as medical science continues to evolve and so will The Healthy Brain Program! In guiding treatments of complex disorders, it is far more advantageous for doctor and patient alike to be broadly right than to be precisely wrong. That kind of insight comes from philosophers like Nicholas Taleb, not from the reductionistic scientific method.

Preventive Medicine: Who owns the brain?

Preventive medicine is a reaction to the 'deficit paradigm'. In regular, allopathic medicine we look at what is wrong and try to fix it. That is what is meant by the 'deficit paradigm' model. Preventive medicine does not look at the patient as a broken widget that can be fixed but looks at environmental factors and tries to optimize them. It began with Dr. John Snow who in the 1850s demonstrated that eliminating a common cause (contaminated water) of illness (cholera) was far more effective than trying to cure the victims. Preventive medicine works; it saves millions of lives at one swoop. Smallpox no longer exists and many diseases, like poliomyelitis are very rare because they have been prevented by proactive means. There is no doubt that prevention can save countless lives often by simple, inexpensive interventions. Aside from public health measures (which are extremely important), methods have been developed to avoid damage and to optimize functions for many organs (heart, liver, pancreas, eyes, etc.). Accurate information and involvement in healthy practices have been shown to positively impact health, life quality and longevity. Perhaps a better understanding of the brain will lead to more respect. After all we have such respect for other aspects of our health. We do not expect people to wear shoes that do not fit. We avoid exposing others to smoke and other toxins. Why do we ignore the needs of the brain?

Interestingly, very little has been done to teach people about the brain as an organ of the body. Why is this topic not approached in elementary schools? Almost nothing is available in an organized, scientifically proven, aesthetically acceptable manner for avoiding brain damage and optimizing brain functions – even in shools and higher educational institutions. We live as if the brain was not nearly as important as other organs. I have had people tell me, after a Healthy Brain Program workshop, that they had lived as if they did not even have a brain! They meant that they were not aware of having a brain, as an organ of the body, with specific vulnerabilities and legitimate health needs.

As I tried to emphasize in the last chapter, physicians often treat common disorders without realizing that they are treating major risk factors for brain disease. They do not educate patients about this – the brain remains a tabu. Generally they do not practice what may be called 'applied preventive medicine'. Is that evidence based?

You are the owner of your own brain, not your doctor. It behooves you to cultivate a sense of proprietorship for this truly marvelous organ of the body.

The Future

Most interestingly, the study of genetics tells us that we have a set of genes that determine our appearance and even the diseases we are susceptible to. This accounts for about 3% of what we are and what happens to us. Our whole genetic code could fit on a CD of about 800 megabytes. Although that is a large amount of information, it clearly isn't enough to dictate everything that happens to us. The science of epigenetics explores how experiences and the environment that we are exposed to activate or unlock certain genetic programs and how they can actually sculpt genes by feeding backwards to alter their influence. There is empirical evidence for this. For example, identical twins who have exactly the same genetic compliment are not nearly 100 % concordant (more like 50%) for many psychiatric conditions, even for the ones that are thought to be highly genetic. These include psychiatric illnesses such as schizophrenia, bipolar illness, autism and attention deficit hyperactivity disorder. Clearly variations in gene expression are at play. Variations are influenced by both the internal and external environments. There are specific chemical reactions that can silence certain parts of the genetic code (DNA) or others that will promote the expression of specific genes. Then there are mutations that also enter the picture, after the code has been established. Complex, is it not? Studies that help us understand the importance of epigenetic influences are vital to the future of brain care. Our experiences interact with our genetic makeup and vulnerabilities to produce either a healthy brain and healthy person who is resilient during adversity, or a person who may succumb to physical and psychiatric illness. These epigenetic influences explain why some twins and siblings do not have the same illnesses and why some twins get the same disease but at different times and at different stages. External factors and internal factors both affect the outcome. Having a genetic program that puts an individual at risk is not always sufficient to produce a disorder or disease (but such conditions do exist). The brain, which is the most plastic (changeable) of all organs, can greatly adapt to and reap the benefits of a beneficial environment and vice versa. Our brain can create an environment that is good for itself. Experience and the passage of time inevitably shape our DNA by silencing some genes and promoting the expression of others, thereby facilitating the cognitive, emotional and behavioural changes that either improve or detract from our makeup (phenotype and endophenotype) and the very quality of life. Of course there will always be people who blame their genes for their difficulties and to some extent they are right. Yet it is not hard to imagine that if you take a pair of identical twins, give one a healthy lifestyle and subject the other one to a sedentary, stressful, unhealthy lifestyle and environment, the two twins will have entirely different health profiles and health spans, not to mention the quality of their lives. That is the essence of the brain-based health approach. With something as complex as the human brain, it will always be easier to prevent something than to fix it. You may quote me on that.

There has been a surprising paucity of scientific studies on what helps the brain. Some are beginning to appear. A recent one from the University of California published in the *American Journal of Geriatric Psychiatry* showed brain changes on an MRI scan after a mere 14-day long healthy lifestyle program.[4] Astounding! More studies like this one will pave the way to effective health promotion, education and prevention of brain diseases.

Biology is about to have its 'big bang' comparable to what happened in physics in the 1930s. This will usher in breakthroughs undreamed of in the past. I will heed the dictum of the baseball great, Yogi Berra, "It's tough to make predictions especially about the future," and I will refrain from the medical habit of prognosticating. There is little doubt however that early diagnosis will become more important as we develop more effective treatment methods. One could say that, in the past, the diagnosis of dementia (and pre-dementia) was unimportant because we understood so little, we had no treatments available and we believed that it was inevitable. That has changed already and it will improve. Today, according to the new classification of psychiatric disease (DSM-V), we have eliminated 'dementia' as a meaningful clinical term![5] Why? Because neuropsychiatrists have become able to image the brain more accurately, in real time, safely in living subjects and they can pinpoint areas of dysfunction. Now we speak of *Major Neurocognitive Impairment* and we include localization and causation in the diagnosis. The imaging technology is fairly new and it dictates that global terms like 'dementia' be abandoned in favor of more accurate descriptors which describe the dysfunction and its anatomical location. Advances in ribonucleic acid chemistry, the characterization and synthesis of neurosteroids and a deeper understanding of epigenetic factors will play a large role in brain disease prevention and future medical care. The development of new pharmaceuticals to better treat chronic inflammation in the brain and the timely use of such drugs may become routine.

As the weakest link in organ repair and replacement, the brain will need all the help it can get. The medical sciences will remain the enchantress of the brain. Future medical technology, sufficiently advanced, will be indistinguishable from magic. We await magic!

GLOSSARY

Abdominal fat: the fat around the midriff and also inside the abdomen surrounding the viscera. Also called visceral fat. It is metabolically more active than the rest of body fat. It is accumulated but also lost at a faster rate.

Abdominal obesity syndrome: a metabolic imbalance with multifactorial causation (genetics, stress, diet, etc) resulting in a pre-diabetic condition characterized by high abdominal fat deposition (high waist circumference). Also known as over-nutrition syndrome, syndrome-x, metabolic syndrome and Raven's disease.

Acetylcholine: a widely distributed neurotransmitter acting in the brain and muscles. It mediates cognitive functions amongst several other vital processes.

Acetylcholinesterase (AChE): an enzyme which degrades acetylcholine.

Acetylcholinesterase inhibitor (AChEI) drug: a group of pharmaceuticals that increase the available acetylcholine in the brain to boost cognitive functions. Compounds already on the market include donepezil, rivastigmine, and galantamine.

Acromegaly: a disease characterized by deforming bone and connective tissue growth caused by excessive somatotropin (growth hormone).

ADAM syndrome: androgen deficiency in the aging male syndrome is characterized by low energy, low libido and erectile dysfunction. May be misdiagnosed as depression.

Adaptation phase: the second phase of the stress response characterized by physiological changes that enable the body to endure the stress. This may actually become an irreversible disease state like diabetes, hypertension etc.

Addiction: the state of being dependent on one or more of the addictive substances. Characterized by craving and unpleasant withdrawal phenomena.

Addictive substances: about ninety substances exist which are addictive because they stimulate the dopamine reward-pleasure pathways in the brain.

Addison's syndrome: a syndrome named after Dr. Addison who was the first to describe a condition caused by low cortisol levels. This syndrome is characterized by low stress tolerance, poor stamina, fatigue and inflammatory states causing muscle and other pains.

Adrenaline: same as epinephrine. A neurotransmitter secreted by the adrenal gland when stimulated by adrenocorticotrophic hormone from the pituitary gland. It has a number of actions on various organ systems to prepare for "fight or flight". An alerting and energizing chemical messenger.

Adult diabetes: diabetes mellitus type II, to be differentiated from type I, the genetic type which is familial and starts in childhood. Unfortunately though we are seeing more and more type II diabetes in children.

Alarm response: the initial phase of the stress response characterized by elevated dopamine, adrenaline and noradrenaline levels in the blood.

Alcohol dehydrogenase: a liver enzyme that degrades alcohol.

Allostatic load: stress load; stressors. The sum of various stressors.

Alpha blocking: the cessation of alpha waves on the EEG when blocked by sensory inputs (eyes open).

Alpha waves, alpha state, alpha rhythm: a calm waking mental state which follows when external auditory and visual (and other) stimulations are reduced or eliminated. The EEG is characterized by 8-13 cycle per second waves.

Alpha-linoleic acid: an essential fatty acid.

Alpha-tocopherol: one of several members of the tocopherol family of antioxidants. The commercially most available form of vitamin E. Natural vitamin E contains many tocopherols and is aptly called mixed tocopherols.

Alzheimer's Disease: a neurodegenerative disease characterized by plaques of amyloid and neurofibrillary tangles that proliferate as the brain is destroyed. On the average, it kills the victim in about ten years. It was first described by Dr. Alois Alzheimer (November 4, 1906).

Amino Acids: organic molecules that are the building blocks of proteins. Although there are hundreds of amino acids in nature, only twenty occur in proteins in humans. Ten of these are essential amino acids that cannot be synthesized by the body and must be taken from food sources.

Amotivational syndrome: a syndrome characterized by a lack of motivation by ordinary rewards and pleasures. It is usually associated with drug abuse but may also be the result of apathy from brain damage or depression.

Amphetamines: a class of strongly stimulating drugs which cause the release of catecholamines.

Amygdala: bilateral structures deep in the brain mediating episodic memory that is related to significant emotional events. They activate the anxiety and "fight-flight" response whenever memories, images or situations signify danger. Discharge of the amygdala in the absence of action is associated with a subjective sense of anxiety.

Amyloidosis, amyloid beta-peptide (A-beta): amyloid beta-peptide is a precursor to the formation of amyloid in the brain. It is believed that this pinkish, gummy, proteinaceous substance is formed in reaction to injury to brain cells. As an inherited trait or gene mutation, some people deposit excessive amounts leading to amyloidosis of the brain which may be the main mechanism in Alzheimer's disease.

Amyotrophic lateral sclerosis (ALS). Lou Gehrig's disease: a degenerative disease of the spine leading to loss of motor function and much later neurocognitive impairment (dementia).

Anabolic process, anabolic hormones: see metabolism.

Androgen (testosterone): the male sex hormone that induces the development of male characteristics including behaviours. It is anabolic to tissues, particularly bones and muscles.

Andropause: a severe drop in testosterone levels; male menopause.

Anemia: low hemoglobin content in blood resulting in a diminished capacity to carry oxygen molecules.

Angina: pain associated with cardiac ischemia.

Angiogenesis: the formation of new blood vessels, capillaries and arterioles, controlled by chemical messengers called brain derived vascular growth factors.

Animal fat: dietary fat from animal sources; high in waxy cholesterol content.

Anorexia nervosa: a neuropsychiatric disorder characterized by severe loss of appetite, food intake and weight.

Anticholinergic effect: effects of herbal preparations or drugs which counteract acetylcholine. Results: dry mouth, dilated pupils, increased heart rate and impairments in attention, concentration and memory. This can progress to psychosis, delirium and death.

Antidepressants: a family of pharmaceutical compounds used to treat depression.

Antidiuretic hormone: see vasopressin.

Antioxidants: a complex and diverse group of molecules that protect key biologic sites from oxidative damage. They act by removing or inactivating free radicals and other reactive oxygen intermediaries. Antioxidant status may be improved through regular exercise(fitness) and dietary measures. Oxidative stress may be induced by toxins (alcohol), infection and other biochemical insults on the body or its cellular components.

Anxiety disorders: neuropsychiatric disorders characterized by high anxiety levels.

Anxiety: activation and unpleasant emotional tension with irrational feelings of gloom and helplessness.

Apolipoprotein E, ApoE gene (Alzheimer's gene): a region of the chromosomes regulating enzymes that control the formation, deposition and clearing of amyloid in the brain.

Artificial intelligence (AI): attempts to model brain functions using computers.

Ascending reticular activating system (RAS): a diffuse network of nerve tissue ascending, supplying incoming information to activate the brain.

Ascorbic acid: see vitamin C.

Aspartate: a neurotransmitter, more accurately called N-methyl-D-aspartate, is found in the brain mediating firing of the nerve cell.

Aspirin: proprietary name for acetylsalicylic acid, a non-steroidal anti-inflammatory drug. Also see salicylate effect.

Atheroma, atheromata: arteriosclerosis. Deposition of fatty, waxy calcium rich plaques usually in arteries and arterioles.

Atherosclerosis: a disorder of arteries characterized by chronic inflammation of the inside of the walls leading to depositions of fat (cholesterol), calcium and clots. This eventually causes hardening of the arteries and even total occlusion leading to ischemia of target tissues.

Atrazine: an herbicide used globally to stop pre- and post-emergence of broadleaf and grassy weeds in major crops.

Atrial fibrillation: a cardiac condition where the atrium, the first chamber of the heart that normally triggers the ventricular contractions, loses its rhythm. Fast erratic signals cause the ventricles to fire too fast or unevenly. The uneven flow through the atrium predisposes to blood clot formation. The danger is that small clots will be pumped to the brain causing silent strokes.

Atrophy: loss of bulk and functional capacity of a tissue usually from lack of use or lack of nutritional or hormonal support.

Attention deficit hyperactivity disorders (ADHD): a neuropsychiatric disorder characterized by age-inappropriate problems with attention, impulse control and hyperactivity.

Autoantibodies and autoimmune disorder: complex molecules are made by the immune system to attach to invading molecules that may cause disease. This process enables the immune system to clear out tagged substances. If antibodies are made to attack the hosts own tissue components, the result is an autoimmune disorder, often difficult to control.

Autonomic nervous system, autonomic functions: part of the central nervous system acting automatically, in a reflex fashion, without awareness or conscious control to regulate sympathetic and parasympathetic output. Sympathetic activity is mediated by adrenaline and noradrenaline and is instrumental in raising heart rate, blood pressure, blood flow to muscles and blood glucose. That is the "fight or flight" response. The parasympathetic system is mediated by acetylcholine which in some ways opposes the sympathetic system. During parasympathetic activity the blood is shunted to the gut and other viscera. Heart rate and blood pressure are lower. Cortisol tends to drop and a sense of relaxation is experienced. This is the "relaxation response" or the "lie down and be loved" response.

Autonomy: desire and ability to manage one's own life. Self regulation.

Axons: a long branch of a nerve cell with terminal synaptic buttons that act to release neuro-transmitters when the axon is fired. The axon communicates with dendrites of other nerve cells.

Bankruptcy hypothesis of aging: a pessimistic demographic and economic projection which forecasts that there will be more and more elders with very few young people to service them with knowledge, skills and tax base.

Beriberi: vitamin B deficiency seen in places where polished rice replaced whole grains. Beriberi is characterized by inflamed nerves, brain disorder and anemia.

Beta-amyloid precursor protein (beta-APP): a precursor protein like substance destined to become amyloid. It is believed that beta-APP is itself toxic to neurons causing some inflammation and destruction at the cellular level.

Beta-carotene: one member of the huge carotenoid family of antioxidants found in coloured vegetables and fruit.

Beta-secretase: an amyloid precursor protein (APP) cleaving enzyme. Drugs to block this enzyme in theory would prevent the build up of beta-amyloid and may help slow or stop Alzheimer's disease.

Biogenic amines: an old term for catecholamines.

Biological clock, body clock: a part of the brain that acts as a timer to pulse other rhythms such as the circadian day-night cycles of hormonal secretions. Many other important cycles are also exist, including: infradian rhythms, which are long-term cycles, such as the annual migration or reproduction cycles found in certain animals or the human menstrual cycle. Ultradian rhythms, which are short cycles, such as the 90-minute REM cycle, the 4 hour nasal cycle, or the 3 hour cycle of growth hormone production. They have periods of less than 24 hours.

Blood pressure disorders: blood pressure is set by the hypothalamus of the brain to remain within certain limits continue to perfuse the brain with blood. If the circulation is poor, blood pressure has to go up. High blood pressure may damage the arterioles causing small leaks and bleeding. Blood pressure that is too low will not propel the blood to the small arterioles which may then plug up. The result is ischemic damage. Unsteady or fluctuating blood pressure may predispose to both types of damage in brain tissue.

Blood sugar: the concentration of blood sugar, more accurately blood glucose, is monitored by the hypothalamic area of the brain and is kept within strict limits. Too much blood glucose, in conjunction blood fat, causes blood vessel lining damages (see atherosclerosis). If the concentration of glucose is too low, symptoms of hypoglycemia appear (hunger, craving, lightheadedness, anxiety, convulsions and even nerve cell death).

Body mass index (BMI): a numeric index derived from height and weight that is a more accurate way to measure overweight and obesity than weight alone.

Bradykinesia: a movement disorder characterized by slowing of muscular movements.

Brain atrophy: loss of nerve and glial tissue with resultant shrinking of the brain.

Brain cells: nerve cells called neurons and feeder cells called glia.

Brain Reserve: see cognitive reserve.

Brain stimulation (auditory, visual, olfactory, etc.): stimulation is important for the brain to keep it functioning optimally. Over-stimulation is stressful and under-stimulation is stupefying. Novel stimuli are associated with learning. Various types of input generate synaptic activity and growth in corresponding areas mapped onto the cortex.

Brain: the central command post of the whole body and seat of consciousness.

Burnout: a state of vital exhaustion.

Calciferol (vitamin D): a fat soluble steroid like substance made by the skin by the action of sunlight. It is a hormone, not a vitamin. Only 5% of optimal requirement comes from dietary sources (fortified milk, cheese and liver oils). It is an anabolic hormone supporting cell differentiation.

Calcium / magnesium: elements required for neuronal function. Large reservoirs are in the bones.

Caloric restriction (CR): restriction of caloric intake to semi-starvation levels. Animal experiments suggest that longevity is facilitated by CR but there is no evidence from human studies.

Calories: see nutritionist calories.

Cannabinoid-like fatty acids: fatty acid molecules found in foods that have an affinity for cannabinoid receptors in the central nervous system.

Cannabis: the hemp plant family which has subspecies with varying amounts of cannabinols that act on cannabinol receptors.

Cardiac arrhythmia: abnormal rate or rhythm of the heart beat.

Carotenoids: a family of thousands of antioxidants called carotenes that are found in colorful fruit and vegetables (like carrots). Beta-carotene is just one of these.

CAT scan, CT scan: computed axial tomography.

Catabolic activity: see metabolism.

Catechins: catechins are polyphenolic antioxidant plant metabolites, specifically flavonoids called flavan-3-ols. Although present in numerous plant species, the largest source in the human diet is from various teas derived from the tea-plant *Camellia sinensis*.

Catecholamines: a family of neurotransmitters which include adrenaline, noradrenaline, dopamine and serotonin. These are involved in regulation of some metabolic activities, autonomic functions, energy level, vigilance and mood.

Cell membrane: a double layer of phosphate and lipid containing structures involved in the transport of molecules across them. It is an intriguing observation (clinical significance unknown) that the nerve cell (neuron) membrane composition reflects the dietary fatty acid profile. Fish oils have high omega-3 content and vegetable oils (corn oil) have high omega-6 fatty acid content. People who have high corn oil consumption (Americans) have high omeg-6/omega-3 ratios in their nerve cell membranes. People who have high fish consumption (Japanese) have the reverse, higher omega-3/omega-6 ratios.

Cerebral edema: swelling of the brain. This is a very serious condition because the brain has no room to swell. The swelling causes increased pressure inside the bony skull which causes further brain injury and swelling leading to a vicious cycle ending in convulsions and death.

Cerebral haemorrhage: bleeding into the soft brain tissue. A type of stroke.

Cerebrospinal fluid (CSF): a clear fluid bathing and nourishing the brain and spinal cord.

Cerebrovascular accident (CVA): A disruption to the blood supply of to a particular region of the brain. Depending on the amount of damage, there may be permanent deficit but usually there is some degree of recovery. A massive stroke or ischemia in sensitive brain areas may cause sudden death.

Cerebrovascular disease (CVD): a disease process, usually inflammatory in nature, leading to occlusion of the arterioles and even larger arteries in the brain.

CHAOS Syndrome: cardiac disease, hypertension, adult diabetes, obesity and stroke.

Chaos: the opposite of order; changing patterns. See nonlinear dynamics.

Chaotic Systems: see complex dynamic systems.

Chemosensory system: hypothalamic regions that monitor the chemical composition of blood and signal for changes to maintain homeostasis.

Cholera: syndrome of severe diarrhea and vomiting. Usually cholera is caused by sewage contaminated drinking water.

Cholesterol: cholesterol is a fat-like pearly substance. It is chemically related to fats and steroids found in animal fats and some vegetable oils. Some organs and tissues are rich in cholesterol (brain, blood, adrenal glands, liver). It is transported in the blood. When its levels are high it may form gallstones in the gallbladder or atheromata on blood vessel walls.

Cholinergic: pertaining to the action of the neurotransmitter acetylcholine, as in cholinergic transmission.

Chronic gum infection: see periodontal disease, gingivitis.

Chronobiology: the study of biological rhythms and their expressions in hormonal and other biochemical activities.

Cimetidine: a drug that inhibits the production of acid in the stomach.

CIND Syndrome: cognitive impairment, no dementia syndrome. Also known as mild cognitive impairment (MCI).

Circadian rhythm: day-night cycle.

Coagulopathies: disorders causing increased viscosity and clotting tendency of blood.

Cognition, cognitive functions: brain functions pertaining to attention, concentration, thinking, memory and executive (planning and decision making) functions.

Cognitive rehabilitation: improvement of cognitive functions by therapeutic and remedial measures following head injury or stroke.

Cognitive reserve: the ability of the brain to recruit previously less used parts, to form new synapses and to rebuild connections. Supported by high synaptic density and high antioxidant status. Brain reserve. Resilience.

Complex dynamic systems: Chaotic systems. Also called non-linear dynamic systems. These complex systems, like the weather systems, ecosystems and oceanic currents, to name a few, are so vastly complex and interactive that they cannot be characterized by usual mathematical models.

Concussion: mild traumatic brain injury (MTBI).

Coronary artery disease (CAD): a disease of the ring of arteries around the heart, usually inflammatory in nature. When one of the coronary arteries becomes occluded, a heart attach follows.

Cortisol: the stress hormone secreted by the adrenal gland. It is a catabolic hormone.

Coup-contrecoup: means "blow counter-blow". It describes what happens to the brain after a blow to the head. As the brain accelerates and decelerates it bounces off the skull. The bruises appear at the point of impact and on the opposite side because of this bouncing effect. One blow yields two lesions.

Cushing syndrome (Cushing's disease): an endocrine disorder caused by elevated cortisol levels. The cortisol usually comes from a tumour but some cortisol like medications can induce the same. The syndrome is characterized by abdominal obesity, hypertension, diabetes and protein loss.

Cyanocobalmin (B12): is a B vitamin essential for blood cells and nervous system functions. Deficiency causes brain and nerve tissue malfunction and anemia.

Cyclic AMP response element binding protein (CREB): proteins that are transcription factors which bind to certain sequences in the DNA on the chromosome thereby increasing or decreasing the transcription of certain genes.

Deconditioned state: lack of fitness because of disuse.

Dementia pugilistica: boxer's dementia. This is a type of dementias from repeated concussions.

Dementia: a brain disorder characterized by memory loss and cognitive deficits. It is a global, somewhat stigmatizing term recently eliminated from the DSM-5 classification of mental disorders. The new term is Neurocognitive Impairment; mild, moderate or severe.

Demographics: the study of population shifts and trends.

Dendrites, dendritic branching: branches of the nerve cell, numbering in many thousands, sending signals to other nerve cells to fire their axons.

Dentate gyrus: part of the hippocampal formation. The dentate gyrus is also one of the few regions of the brain where neurogenesis takes place. Neurogenesis is thought to play a role in the formation of new memories. It has also been found to increase in response to both antidepressants and physical exercise. Conversely, endogenous and exogenous glucocorticoids (stress hormones) such as cortisol inhibit neurogenesis. Both endogenous and exogenous glucocorticoids are known to cause psychosis and depression, implying that neurogenesis may improve symptoms of depression.

Depression, depressive illness: a mood disorder characterized by low energy or agitation, anxiety and feelings of hopelessness.

Desoxyribonucleic acid (DNA): A complex spiral molecule containing a code made of sequences of four simpler chemicals, the genetic code, carried in the chromosomes.

Diabetes: see adult diabetes.

Dichlorodiphenyltrichloroethane (DDT): a pesticide that has been banned in North America. It is an estrogen mimic that is still found in the food chain.

Diethylstilbestrol (DES): a synthetic estrogen with harmful side effects on the foetus.

Dioxin: a highly toxic and carcinogenic industrial by-product of incineration. There is some concern because it has found its way into the food chain.

Diplopia: double vision.

DNA: see desoxyribonucleic acid.

Docosahexaenoic acid (DHA): trivial name cervonic acid, is an omega-3 essential fatty acid. Chemically, DHA is a carboxylic acid with a 22-carbon chain. DHA is most often found in fish oil. Most of the DHA in fish and other more complex organisms originates in microalgae of the genus *Schizochytrium*, and concentrates in organisms as it moves up the food chain. Most animals make very little DHA metabolically, however small amounts are manufactured internally through the consumption of alpha-linoleic acid, an omega-3 fatty acid found in chia, flax, and many other seeds and nuts. DHA is a major fatty acid in sperm and brain phospholipids, and especially in the retina. Dietary DHA can reduce the level of blood triglycerides in humans, which may reduce the risk of heart disease. Low levels of DHA cause reduction of brain serotonin levels and have been associated with ADHD, Alzheimer's disease, and depression, among other diseases, and there is mounting evidence that DHA supplementation may be effective in combating such diseases.

Dopamine: a catecholamine neurotransmitter acting in the pleasure and reward system pathways. It also modulates involuntary muscle movements.

Dreams, dreaming: see rapid eye movement REM sleep.

Dualistic thinking: using two models where one cannot adequately explain a phenomenon. The brain can be thought of as a computer and an endocrine gland. There is a wave theory and a particle theory to explain the physics of light.

Dyslipidemia: a condition marked by abnormal concentrations of lipids or lipoproteins in the blood.

Eating disorders: neuropsychiatric disorders resulting in disturbances in the regulation of appetite and body mass.

Electroencephalogram (EEG): a graphic representation of the electrical activity of the brain by displaying the waves picked up by a number (usually 12) electrodes place on the scalp.

Electron microscopy: a form of microscopy using electron beams instead of light. Much higher magnifications can be reached enabling observations of intracellular structures.

Embolism: the occlusion of a blood vessel by a plug of some substance, commonly a blood clot or atheroma (the waxy plaque deposited on the artery wall as a result of atherosclerosis).

Emotional brain: The brain is modular and has specific regions. The emotional centre and reward centre are related and are called the limbic system. The limbic system, also called the Papez circuit, exists as a double circuit comprised of the olfactory bulb and tract, the cingulate gyrus of the cortex, the fornix and hippocampus and amygdala. The limbic system mediates motivation, pleasure, mood and also influences autonomic functions via the hypothalamus.

Endocrine system: a number of glandular organs systems which secrete chemical messengers (hormones) into the blood or tissues to influence distant organs or structures. Accordingly, the brain can be conceived as an endocrine gland.

Endorphins: a family of hormones and hormone like substances resembling opiates and acting on internal opiate receptors to modulate feelings of well being and pain. Internal natural opiates.

Energy: The form of energy that fuels cells is stored in the molecule adenosine 5-triphosphate (ATP) which a multifunctional nucleotide that is most important as a "molecular currency" of intracellular energy transfer. In this role ATP transports chemical energy within cells for metabolism. It is produced as an energy source during the processes of photosynthesis and cellular respiration and consumed by many enzymes and a multitude of cellular processes including biosynthetic reactions, motility and cell division. ATP is also incorporated into nucleic acids by polymerases in the processes of DNA replication and transcription. The formation of ATP requires energy and its degradation is accompanied by a release of energy.

Enrichment: supplying a variety of novel and healthy stimuli to the brain to enable learning. This results in an increase in synaptic density.

Environmental deprivation: reduction of a healthy variety of stimuli to the brain that may lead to loss of synaptic density. Opposite of enrichment.

Enzymes: organic molecules that act to promote or inhibit a biochemical reaction.

Epidemic: in epidemiology, a disease that appears as new cases in a given human population, during a given period, at a rate that substantially exceeds what is "expected", based on recent experience (the number of new cases in the population during a specified period of time is called the "incidence rate"). See epidemiology.

Epidemiology: the study of diseases that afflict populations. This involves the measurement of new cases (incidence) as well as an estimate of existing cases (prevalence). Epidemiologists are interested in how disease patterns shift and what trends are emerging.

Epigenetics: the study of experiential and environmental factors that may result in inhibiting or expressing an inherited genetic program.

Epinephrine: see adrenaline.

Essential amino acids: indispensable amino acids, are amino acids that cannot be synthesized *de novo* by a human and therefore must be supplied in the diet. Nutritional essentiality is characteristic of the species, not the nutrient. Ten amino acids are generally regarded as essential for humans. They are: histidine, isoleucine, leucine, lysine, methionine, phenylalanine, threonine, tryptophan, arginine and valine. In addition, the amino acids cysteine, glycine and tyrosine are considered conditionally essential, meaning they are not normally required in the diet, but must be supplied exogenously to specific populations that do not synthesize it in adequate amounts.

Essential fatty acid deficiency (EFAD): a syndrome characterized by the suppression of the inflammatory and other immune system functions.

Essential fatty acids (EFA): fatty acids that are required in the human diet; they must be obtained from food as human cells have no biochemical pathways capable of producing them internally. There are two families of EFAs: ω-3 (or omega-3 or n-3) and ω-6 (omega-6, n-6.) Fats from each of these families are essential, as the body can convert one omega-3 to another omega-3, for example, but cannot create an omega-3 from scratch. They were originally designated as Vitamin F when they were discovered as essential nutrients in 1923. In 1930, work by Burr, Burr and Miller showed that they are better classified with the fats than with the vitamins.

Essential nutrients: Under some circumstances particular foods will prevent disease. When no other sources of vitamin C were available, limes prevented scurvy in British sailors (called "limeys"). Unpolished, natural rice prevents beriberi (a vitamin B deficiency disease). Liver prevents night blindness, and so froth. The human body can synthesize everything it needs from a small number of substances available through food and sunshine. However, there are a number of chemical substances that it must have to succeed at this very complex but vital process. Such essential nutrients must come from food as they are required for normal body functioning and cannot be synthesized by the body. Categories of essential nutrients include vitamins, dietary minerals, essential fatty acids and essential amino acids. Different species have very different essential nutrient requirements.

Estradiol: the human form of the estrogen family of hormones.

Estrogen replacement therapy (ERT): the replacement of estrogen in estrogen deficiency caused by the loss of ovarian function, as after surgical removal or menopause.

Estrogen: the female sex hormone. Actually a family of molecules both natural and synthetic that interact with estrogen receptors in the brain and other estrogen sensitive tissues.

Eustress: optimal stress, good stress.

Evidence based medicine: the examination and understanding of medical information according to levels of evidence. Anecdotal report are the lowest.

Prospective, double blind randomized trials using control comparison groups and meta-analyses are the highest.

Executive functions: the highest functions of the brain. Mediated in the prefrontal cortex. Executive functions pertain to the ability to judge time, extract meaning, make decisions and plan for the future.

Exhaustion phase: the third and final phase of the stress response characterized by deregulation of the

hypothalamo-pituitary-adrenal axis resulting in low or depleted cortisol levels and corresponding (Addisonian) symptoms.

Fats: greasy fatty/waxy organic substances which do not dissolve in water. Saturated fats: some naturally occurring fats are called saturated because all the carbon atoms have four single bonds. These fats, usually of animal origin, are solid at room temperature. Coconut and palm oils are also saturated. Heart disease, vascular disease and cancer are associated with high saturated fat diets. Unsaturated fats: a mix of monounsaturated and polyunsaturated fats. Eating unsaturated fats lowers cholesterol when substituted for saturated fats. Polyunsaturated fats: fats and fatty acids with more than one double bond on the carbon atom in the chain. They are soft at room temperature. These include corn, sunflower, safflower and canola oils. Monounsaturated fats have only one double bond in their carbon chain. They are found in nuts, peanuts, avocado and olives and their oils. Trans-fats: hydrogenation is a process that adds hydrogen to an oil to make it more solid and stable. It also creates trans-fatty acids which act like saturated fats.

Fatty acids: components of fat that can exist in the free form but usually in compounds called glycerides. Some

fatty acids are called essential fatty acids (linoleic and alpha linoleic acids) because they cannot be made by the body and must be ingested.

Feeding reflex: the complex sequence of motor functions involved in taking in food (biting, chewing, swallowing). Once initiated it tends to go on autonomously until inhibited.

Fibre: components of food that are structural and give it bulk and moisture retaining properties. This is important for bowel function and digestion. Fibre is broken down more slowly in the gut to yield important nutrients, not merely bulk.

Fibrin: A whitish insoluble protein substance formed by a number of clotting factors in blood and tissues. Fibrin forms the essential portion of the blood clot.

Fight or flight response: the activating energizing stress response mediated by the sympathetic autonomic nervous system and specific neurotransmitters and hormones. It prepares the body for strenuous activity and coping with injury. Vigilance and concentration are increased. Blood is shunted to muscles. Heart rate, blood pressure and respiration increase. Glucose enters the blood. Blood clotting tendency also increases.

Fitness: optimal strength, endurance and flexibility.

Flavanoids: a generic term for a group of compounds called bioflavonoids derived from plans (fruits and vegetables) which are involved in the maintenance of healthy blood vessels walls.

Focusing: a meditative technique using imagery and self awareness developed by psychologist, Eugene Gendlin, Chicago.

Folic acid: a B vitamin required for normal growth of cells and optimal neuronal functioning. Deficiency may cause psychiatric symptoms.

Fractals: very complex patterns that derive from a repeating set of simple instructions. James Gleick (author of national bestseller *CHAOS*) said "in the mind's eye, a fractal is a way of seeing infinity".

Free radicals: oxygen atoms (also known as rogue oxygen species) dissociated from oxygen molecules are highly reactive with other molecules causing their destruction by oxidation. Lipid molecules which make up intracellular components and cell membranes are highly susceptible to this process.

Frontal lobes: the foremost and most recently evolved part of the brain mediating inhibition, working memory, cognitive functions, executive and other higher functions (e.g. meaning, insight, empathy). The organ of civilization.

Frontotemporal dementia (FTD): this term pertains to a region of the brain, not the mechanism of disease. One type of FTD is characterized by dysinhibited behaviours, lack of insight impulsivity, another type by apathy and amotivational syndrome.

Gamma-aminobutyric acid (GABA): an inhibitory neurotransmitter widely distributed in the brain, particularly in the cortex.

Gender specificity: the term used in scientific studies to denote findings pertinent to one gender and not the other.

Gene expression: the translation of a genetic code sequence found on the chromosome. This process can be inhibited or triggered by body chemistry related to environmental, epigenetic factors.

General adaptation syndrome (GAS): see adaptation phase.

Genes: chemical (nucleotide) code sequences inheritable via the chromosomes.

Genetic polymorphism: the variety of ways a genetic message can be expressed in various populations or individuals.

Genetics: the study of gene transmission and expression.

Gingivitis: see periodontal disease.

Glia, glial cells: brain cells between neuron cells that are not involved with bioelectric signaling but have an intimate biochemical coupling with the neurons. Neuron feeder cells.

Glucose: a simple sugar molecule.

Grapefruit effect: an effect of grapefruit and some other juices causing increased absorption and slowed degradation of an unrelated drug leading to higher than expected blood levels. This can cause a relative overdose with dangerous effects and/or side effects.

Grey matter of brain: the outermost layer of cerebral cortex, about 3 mm thick, comprised of 6 layers of nerve cells.

Growth hormone (somatotropin): a hormone secreted by the pituitary gland supporting the viability of many tissues by increasing protein anabolic activity, particularly connective tissue and bone matrix.

Habit: relatively autonomous repeated behaviours. It takes about thirty days of practice to establish a habit. Once established the behaviour becomes autonomous or self sustaining, without decision making effort. Inability to carry out the habit gives rise to craving and discomfort. Likewise, it takes time to break a habit.

Health span: the length of time a person is in a state of good health without loss of quality of life.

Healthy heart programs: programs usually in the community or hospitals, dedicated to improving heart health by promoting physical fitness, healthy diet and stress management.

Healthy narcissism: positive self regard and self respect.

Heart attack: acute myocardial infarction.

Hemoglobin: a complex molecule containing iron found in the red cells of blood. Hemoglobin has an affinity for oxygen molecules picking them up in the lungs and carrying them to other tissues.

Hemorrhage: uncontrolled bleeding.

Herbs: plants with spicy or medicinal properties.

High blood pressure: see hypertension.

High blood sugar: see hyperglycemia.

Hippocampus: bilateral structures in the brain mediating encoding of new information (learning) and retrieval of information (memory).

Histamine: a neurotransmitter substance that mediates dilation of capillaries, gastric secretion and wakefulness. It also plays a role in the inflammatory response.

HIV: human immunodeficiency virus.

Homeostasis: the process of maintaining a constant internal environment in the body.

Homocysteine: byproduct of methionine and B vitamin metabolism. Elevated levels correlate with cerebrovascular disease but it is not known whether this is a cause or merely the result (a marker) of a more fundamental disorder. Several pathogenetic mechanisms linking decreased activity of vitamin B12, folic acid and elevated homocysteine to dementia are proposed but none have been proven and it may be that they all act concurrently.

Hormones: chemical messengers secreted by a gland to act on a target tissue specific to that messenger.

Human post-mortem studies: autopsy studies.

Hunger and cravings: a complex response to biochemical imbalances in the blood composition, including but not restricted to glucose, registering as a subjective feeling that there is a need for that substance. Hunger is a response to a true lack of food and energy. Cravings for substances are mostly response to acquired tastes.

Hypercholesterolemia: abnormally high cholesterol levels in the blood.

Hyperglycemia: abnormally high sugar (glucose) levels in blood.

Hyperlipidemia: abnormally high fat (lipid) levels in the blood.

Hyperparathyroidism: abnormally high parathyroid hormone levels in the blood causing high calcium turnover from bone. Shifts in serum calcium levels cause changes in cognition and mood.

Hypertension: abnormally high blood pressure.

Hyperthyroidism: abnormally high levels of thyroid hormone (thyroxin) in the blood.

Hypnagogic hallucinations: hallucinations, mostly visual, just before falling asleep. Probably the early stages of a REM period. This is a rather common phenomenon, not signifying a disease.

Hypnogram: a graphic record of sleep activity including wakefulness, respiration, oxygen saturation and motor activity, but not necessarily sleep EEG.

Hypnotics: drugs that induce sleep.

Hypothalamo-Pituitary-Adrenal (HPA) Axis: the core structures active in mediating the stress response. The hypothalamus takes command from limbic and cortical areas. It releases corticotropin releasing factor (CRF) starting the cascades down to the pituitary gland where corticotropin or adrenocorticotrophic hormone (ACTH) is released to act distally on the adrenal gland which releases cortisol and adrenalin. They in turn act on a number of organ systems preparing them for the "fight or flight" response.

Hypothalamus: area deep in the midbrain subcortical area mediating autonomic endocrine activity.

Hypothyroidism: abnormally low levels of thyroid hormone in blood.

Iatrogenesis: problems resulting from the activity of physicians.

Incidence: see epidemiology.

Infarct, infarction: an area of blood clot and tissue death.

Independent risk factors: risk factors for a disease that have an effect independently of each other and when present together, act in a synergistic, additive manner.

Inflammation: a self protective biological response to tissue injury characterized by increased blood flow, swelling from collection of fluid and clotting and pain. In the brain, the pain is absent.

Inflammatory disease of connective tissues (autoimmune disorders): inflammation secondary to the body's immune system attacking tissues of the self.

Insomnia: inability to fall asleep or stay asleep well enough to get a refreshing sleep.

Insulin, insulin resistance: insulin is a hormone from particular pancreatic cells which acts as a chemical messenger to enable muscle and other cells to absorb glucose. In diabetes, glucose cannot enter the cells efficiently; they are said to be insulin resistant. As a result, both blood glucose and insulin levels increase.

Interactionism: this term refers to the dynamic state of complex, including biological systems. They are on the move, always changing and the changes affect each other.

Iodine deficiency: iodine, in miniscule amounts, is needed by thyroid tissue to form thyroid hormone. Total deficiency of iodine in the diet can cause a type of hypothyroidism.

Iron, iron deficiency: this element is needed by the bone marrow to make hemoglobin for the red blood cells. It is the hemoglobin that gives the blood its red colouration. Deficiency results in a type of anemia called "iron deficiency anemia" which can be corrected by iron supplements. Also see hemoglobin.

Ischemia, ischemic damage: ischemia refers to a state where there is not enough blood to supply oxygen to the surrounding tissue. This oxygen deprivation results in the transient or even permanent loss of function (cell death) in the affected area.

Ischemic heart disease: the heart muscles have a great demand for oxygen. Small losses in blood supply to heart muscles will cause damage to muscle cells resulting in weakening and or irritability. A more complete loss of blood/oxygen supply will cause a frank heart attack, a myocardial infarction.

Ischemic white matter lesions: poor blood supply in the deeper parts of the brain, where the myelin covered nerve tracts are located, will cause small areas to suffer cell loss. These areas are small silent strokes also known as white matter infarctions. An accumulation of these will result in progressive vascular neurocognitive impairment (dementia).

Knock-out genes, knock-out mice: genes that have been deactivated or removed from their chromosomes resulting in an inheritable deficit in the transcription corresponding with that region. The result is an inherited biochemical deficiency expressed as a disease state. Mice (the usual candidates) that are bred with such deficits are called knockout mice. For example, there are knock-out mice that express Alzheimer disease like brain amyloidosis secondary to injuries.

Kwashiorkor: (local name in Gold Coast, Africa "golden boy" or 'red boy'). Severe protein deficiency with characteristic pigmantation of skin and hair. *Marasmic kwashiorkor* is a severe condition with deficiency of both calories and protein leading to severe tissue wasting, loss of fat and dehydration.

Lead poisoning: acute or chronic lead ingestion, usually from paint, resulting in cognitive impairment and anemia.

Leptin: the hormone released by fat cells to signal satiety.

Life expectancy et birth (LEB): the length of time a person is expected to live from the moment of birth, expressed as an average for a given population. Expected life span. Life expectancy after birth may eventually exceed the average life span because some risk factors are eliminated as one ages, especially if one ages in a healthy way.

Life span: life expectancy at birth.

Light Stimulation: the use of bright lights to influence the diurnal rhythm. This may treat some sleep and mood disorders.

Limbic, hypothalamus, pituitary and adrenal gland (LHPA) axis: emotional brain and autonomic nervous system expressing itself through the stress-endocrine system. The limbic system plus the HPA axis.

Linear systems, liner dynamics: simple logical equations. Classical math and physics.

Lipid peroxidases: chemicals that degrade fats by oxidation into simpler components.

Lipids, lipid disorders: lipids are a broad category of oils, fats and waxy substances. Disorders can arise from imbalances in the composition or transport of lipids in blood. Example: too much "bad cholesterol".

Liver enzymes: enzymes found in the liver that break down complex chemicals and toxins.

Long term memory: relatively permanent memory encoded via the hippocampus into various brain regions. Such a memory is re-membered upon recall.

Longitudinal Studies: population studies that extend into the future.

Loss of consciousness (LOC): loss of organized response to external stimuli and loss of awareness of a segment of time and reality.

Lou Gehrig's disease: see amyotrophic lateral sclerosis.

Low density lipoproteins (LDL) (Bad Cholesterol): molecules of fat in the blood that tend to stick to the walls of arterioles, especially if the walls are damaged and blood sugar is elevated.

Lycopenes: red coloured carotenoid pigments with strong antioxidant properties.

Macular degeneration: degeneration of the macula lutea (white spot) of the retina. Leading cause of blindness.

Magnetic Resonance Imaging (MRI): a type of imaging, like x-rays, but using a strong magnetic field to detect and differentiate tissue boundaries.

Manpo-Kei: the Japanese practice of step-counting to increase activity level.

Marasmus: (dying away) atrophy and wasting away without a clear medical cause. May be from caloric and/or nutrient deficiency. *Marasmic kwashiorkor* is a severe condition with deficiency of both calories and protein leading to severe tissue wasting, loss of fat and dehydration. See kwashiorkor.

Mediterranean Diet: a diet rich in fruits, vegetables, sea food and olive oil.

Megabyte: a million bytes. A byte is eight bits of information, like 0 or 1 (yes or no).

Melatonin: hormone of darkness synthesized from serotonin and secreted by the pineal gland.

Memory: capacity for storage of all types of information. Short term or working memory is a cortical function, very dependent on concentration and lasts only for seconds or minutes. More durable memories are formed when the hippocampus processes (dismembers) the 'memory' and allocates it to various brain regions to be 're-membered' again by the hippocampus later, on demand.

Meningiomas: non metastasizing, benign tumours of the brain covering (meninges). They can occur spontaneously or in association with growth hormone use.

Menopause: a cessation of menses resulting from ovarian hormone decline.

Mental exercise, stimulation: any brain activity in various domains. Auditory, visual, tactile, olfactory, kinesthetic and gustatory activities all count as brain stimulation, have specific representations and storage capacity (memories) associated with them.

Meta-analysis: the study of studies.

Metabolism, metabolic rate: metabolism refers to the overall biological chemical processes of a living organism. It proceeds at a particular rate governed, in part, by body temperature. Metabolism is the balance between the anabolic, body building and repairing processes (anabolism) and the catabolic, body degrading and dissolving processes (catabolism). The metabolic processes are regulated by thousands of enzymatic reactions and anabolic and catabolic hormonal effects.

Metamemory: a memory of a memory.

Methionine: a sulfur bearing amino acid essential for optimal growth and nitrogen balance.

Methoxychlor: a pesticide.

Methyldopa: antihypertensive agent. Opposes action of the neurotransmitter dopamine.

Methylxanthines: substances in the xanthine family of compounds. These come from yellow vegetables and are protective of neural, particularly retinal tissue. Lutein and zeoxanthine are prescribed to prevent macular degeneration.

Microvascular inflammation: inflammation, usually chronic, of the small arterioles in various regions. It is believed that microvascular inflammation ushers in arteriosclerosis.

Migraine conditions: symptoms produced by unstable vascular reactions like chronic constriction and dilatation of blood vessels. As a diseas, this commonly happen in the abdomen, eye and brain.

Mild cognitive impairment (MCI): subjective or mild memory impairment without signs of major neurocognitive impairment (dementia).

Mild traumatic brain injury (MTBI): concussion

Mind: the faculty or function of the brain, by which an individual becomes aware of surroundings and space and time and by which one experiences emotions, feelings and desires and is able to attend to, remember, to reason and decide.

Mixed carotenoids: a balanced, not just one member (usually marketed as beta-carotene) of the carotenoid family.

Mixed tocopherols: a balanced, not just one member (usually marketed as alpha-tocopherol) of the tocopherol family.

Morbidity compression: rectangularization of the health span. The application of lifestyle and other preventive measures to promote health well into old age thereby pushing disease and death back to the last few years or months of life. Salutogenesis.

Mortality rate: death rate usually expressed as a number per 100,000 of a given population.

Mozart effect: the relaxing effect of certain types of music exhibiting longer periodicity. Not only Mozart but Handel, Haydn, Bach, Schubert, Vivaldi, Teleman have also written such music.

Multiple sclerosis (MS): the loss of myelin, the insulation around nerves in the brain.

Myocardial infarction: heart attack. Death of a region of heart muscle from inadequate blood supply.

Narcolepsy: sleep attack.

Neuroanatomy: the study of the anatomy of the brain and nerves.

Neurocognitive Impairment (NCI): impairment in higher brain functions pertaining to concentration, memory, speech and/or executive functions. Formerly referred to as 'dementia'.

Neurodegeneration: degeneration and death of nerve cells in the brain, spine or other parts of the body.

Neurofibrillary tangles: microscopic debris of remnants of nerve cell fibres found with plaques in Alzheimer's disease.

Neurogenesis: the birth of new brain cells.

Neuroimaging: various methods like X-rays, CT, MRI and PET scans to create visual images of the brain often. Functional imaging refers to the method, usually MRI, of seeing the brain in real time while it is working.

Neuron: brain or nerve cell.

Neuropeptides: molecules made of amino acids having modulating activity in the functioning of the brain.

Neuroplasticity: the ability of the brain to make new tracts and rewire itself and the ability to recruit new parts to perform previously learned activities.

Neuroprotection: the protective effect on the degradation of neurons by some substances such as antioxidants.

Neuropsychiatric disorders: psychiatric disorders.

Neurosteroids: natural steroids based on the cholesterol molecule having modulating effects on the brain.

Neurotoxins: substances that are toxic specifically to nerve or brain cells.

Neurotransmitters: substances that can excite or inhibit nerve the cell membrane to facilitate or slow its firing.

Nicotine: an alkaloid neurotransmitter with excitatory and inhibitory activity depending on where it acts. Nicotinism is poisoning by nicotine.

Nitrates: nitric acid salts commonly used as food preservatives. Usually well tolerated but can cause symptoms in some people.

N-metyl-D-aspartate (NMDA): an excitatory amino acid neurotransmitter found widely distributed in the brain.

Nonlinear dynamics: the study of complex, ever changing interactive systems.

Non-rapid eye movement sleep (NREM): deep slow wave sleep.

Non-steroidal anti-inflammatory drugs (NSAIDs): pharmaceuticals that are not related to or do not act by mimicking cortisol for anti-inflammatory effect.

Noradrenaline: see norepinephrine.

Norepinepherine: also known as noradrenaline is a neurotransmitter mediating activation of the autonomic nervous system and brain regions. As one example, it elevates blood pressure.

Nutrient-caloric ratio: N/C. An arbitrary nutrient value divided by the number of nutritionist calories consumed. The higher the number the more nutritious is the food. Pure starch, sugar and alcohol may supply a lot of calories but their N/C = 0.

Nutritionist calories (Kcal): In Latin, *calor* means heat. Calorie is a unit of heat energy. The small **c** or small Calorie is used in physics. It is the amount of heat required to raise the temperature of one gram of water by one centigrade degree. In metabolic studies and nutrition, the large calorie or Kcal (1,000 calories = 1 kilocalorie) is used. This is the nutritionist calorie or **C**. Protein = 4 C/gram. Carbohydrates = 4 C/Gram. Fat = 9 C/gram.

Obesity: an increase in body weight beyond the limitations of skeletal and physical requirements as a result of excessive accumulation of fat in the body. Based on the normal BMI range of 20-25 kg/m^2, overweight is above that and clinical obesity starts at 30 kg/m^2.

Obstructive sleep apnea (OSA): periodic cessation of breathing during sleep caused by obstruction of the pharynx usually by the soft palate and surrounding tissues.

Oils: liquid fats.

Omega fatty acids: a polyunsaturated fatty acid family built from linoleic acid. Omega-3 and omega 6 fatty acids are needed in approximately equal amounts to make hormone like substances called eicosanoids that compete for metabolic control of the body. Omega-3-fatty acids have anticoagulant properties. Also see cell membrane.

Opiates: any substance containing or derived from opium.

Optimal daily allowance (ODA). a new concept emerging in reaction to the inadequacies of the recommended daily allowance (RDA) that was derived from an effort to analyze "normal" eating habits in order to develop adequate rations for soldiers in WWII. The RDA is biased towards high proteins and fats, high calories and low nutrients (vitamins, minerals, etc).

Orexins: The hypocretins (orexins) are a newly identified peptide family. Recent observations suggest an involvement of these peptides in producing wakefulness.

Osteoporosis: abnormal thinning and rarefaction of bone due to the failure of bone cells to lay down bone matrix.

Over-nutrition, over-nutrition syndrome: see abdominal obesity syndrome.

Overweight: see obesity.

Oxidation, oxidative stress: see antioxidants.

Oxygen radical absorbing capacity (ORAC): a measure of anti-oxidant capacity.

Oxytocin: a hormone from the pituitary gland. It promotes smooth muscle contractions but in the brain it promotes pair bonding behaviour and higher levels of trust.

Parathormone: parathyroid hormone. Secreted by the parathyroid glands in the neck to regulate extracellular calcium levels. It is regulated by feedback and also magnesium and vitamin D levels. It has effects on brain related to anxiety and mood but neuropsychiatric effects have not been well researched.

Parathyroid Hormone: see parathormone.

Parkinson's disease: also known as *paralysis agitans*. A degenerative brain disorder characterized by mask-like face, tremor and stiffness of muscles, festinating gait and progressive mood and cognitive changes. It may also be secondary to trauma or specific toxins. The brain region affected is the substantia nigra.

Pedometer: a device that can measure steps. Often it can be calibrated to the individual's gait so that the results can be read as steps, kilometers or miles traveled.

Peptide: two or more linked amino acids.

Periodontal disease: gingivitis, gum infection and inflammation.

Peripheral artery disease (PAD): vascular disease, usually atherosclerotic disease of the arteries of the extremeties. Often associated with atherosclerosis of central structures like heart and brain.

Persistent organic pollutants (POPs): synthetic organic compounds produced for industrial purposes, that are relatively inert and do not break down easily. They enter the food chain and are concentrated in older animals that feed on contaminated species. Toxic tissue levels have been in some species causing hormone havoc and even death.

Persistent post-concussion disorder (PPCD): post-concussion symptoms that last for more than two years.

PET Scan: positron emission tomography scan of an organ.

Pharynx: mouth and throat cavities. Area of mucous membrane surrounding glandular tissues and muscles between nasal passages, mouth and esophagus.

Phenotype: outward visible expression of the hereditary constitution (endophenotype = inner biochemical hereditary constitution).

Phospholipids: a class of lipids containing fatty acids, phosphorus and nitrogen. They are important components of cell membranes.

Physical exercise: see fitness.

Phytotherapy: treatment by use of plants.

Pilot study: a small study to ascertain whether a more sophisticated and bigger study would be important or feasible.

Pituary gland: an important gland near the base of the skull at the back of the throat. It is influenced by the brain via the hypothalamus. It orchestrates most of the releasing factors and stimulating factors for the endocrine glands.

Placebo effect: an experimental finding that appears to depend on positive expectations and other psychological factors. The response to an inert substance or simulated procedure. As much as 20-40% of findings may be placebo effects.

Placebo: an inert substance or sham procedure to serve as a control for the active ingredient or genuine procedure.

Platelet clumping: a preliminary stage of blood clotting. Increases blood clotting tendency.

Pleasure pathways of brain, reward centre: the dopamine pathway.

Polychlorinated biphenyls (PCBs): an inert plastic, no longer manufactured but still in transformers, electrical components and the food chain. In high concentrations and long exposures it is toxic to a variety of tissues and biochemical systems.

Polypharmacy: the use of multiple pharmaceutical compounds. Polypharmacy may increase the risk of side effects and drug interactions.

Polyphenols: antioxidant and ant-cancer agents commonly found in many plants, particularly green tea.

Population projections: forecasting of population trends based on epidemiological studies.

Positive thinking, power of: positive thoughts in the cortex can induce good feelings in the limbic system and viceversa.

Post traumatic stress disorder (PTSD): a neuropsychiatric syndrome caused by an extremely stressful event. It is characterized by sleep disorder, anxiety states, panic attacks and memory intrusions (flashbacks) that may be triggered by particular events.

Post-concussion disorder (PCD), post-concussion syndrome (PCS): symptoms associated with concussion (aka mild traumatic brain injury = MTBI). These may include headaches, anxiety, sleep disorder, mood changes, fatigue and problems with attention, concentration and memory.

Post-traumatic amnesia (PTA): a period of changed consciousness immediately after a concussion (aka mild traumatic brain injury = MTBI).

Prefrontal cortex: the tip of the frontal cortex and frontal lobes.

Prevalence: see epidemiology.

Preventive medicine: the application of biological and medical knowledge to prevent diseases, usually in large numbers of a given population.

Progesterone: the hormone of gestation secreted by the ovaries. It has effects on the uterus and calming emotional effects.

Prolactin: the hormone of lactation. It has affects the breast tissues, suppresses ovulation and has a calming emotional effect.

Protein: derived from the Greek *proteus* meaning primary. Proteins are the main building blocks of the body. They are complex molecules made up of chains of amino acids. Proteins form the major constituents of muscle, catalyze some chemical reactions, regulate gene expression and compose major structural elements of all cells.

Psychoneuroendocrinology: the study of hormonal effects on brain functions.

Psychopharmacology: the study of the effects of pharmaceuticals on brain functions.

Psychosis: severe derangement of brain function altering reality sense, perceptions and communication to the extent that they impair the ability to survive.

Rancidification: the process of putrefaction and decomposition of fats resulting in a musty, rank taste or smell from the liberation of fatty acids.

Randomized, placebo controlled trial (RCT): see scientific method.

Rapid eye movement (REM) sleep: a stage of sleep characterized by paralysis of muscles but not eye movements. The EEG shows activation all over the cortex. Dreaming takes place during REM sleep.

Reaction time: short time lapse between a perception and a required action.

Receptors: molecular complexes on cell membranes that react to specific substances and alter the polarization of the membrane.

Recommended daily allowance (RDA): see optimal daily allowance (ODA).

Rectangularization of the health span: see morbidity compression.

Relaxation training: methods of attaining a relaxed state with lowered stress and stress hormone levels. These methods, self taught or taught by a trainer, often use imagery, meditation, deep breathing and mild physical exercises to approach an *alpha state* or at least a relaxed pleasant state of mind. When this becomes cued and follows as an automatic response, it is the opposite of the "fight-flight" response and has been called the "relaxation response".

REM associated behaviour disorder (RBD): see sleep paralysis.

Restless Leg Syndrome (RLS): an unpleasant sensation of having to move the legs. Occurs as a sleep disorder.

Resveratrol: a powerful antioxidant giving a reddish blue colour as found in blueberries, red wine, etc.

Retinal hemorrhages: bleeding into the retina of the eye.

Reward system: the pleasure pathway of he limbic system mediated by dopamine.

Ribonucleic acid (RNA): high molecular weight compounds that are made up of simpler building blocks called nucleotides. Nucleotides, occurring in specific sequences carry information transcribed from the genetic code. RNA is s molecule that can carry a message.

Rogue oxygen species: see free radicals.

Runner's high: feeling of well being and even euphoria as a result of internal opiates (endorphins) and large amounts of catecholamines released during prolongrd heavy exercise.

Salicylate effect: the pharmaceutical property of acetylsalicylic acid (ASA) which is to decrease the stickiness of blood components causing lower blood viscosity and less clotting tendency. Many herbal compounds contain salicylate and those have additive effects.

Schizophrenic illness: a psychosis characterized by delusions, hallucinations and thought disorder.

Scientific method: the derivation of knowledge by the testing of a hypothesis by means of carefully controlled experiments or studies. The comparisons are made against a control group that is not part of the experiment. Attempts are made to reduce bias by randomizing members of the control group and by using placebos. Hence, randomized controlled trial (RCT) that is the gold standard of the scientific method. The scientific method does not prove anything but disproves specific assumptions.

Second impact syndrome: the result of a simple concussion may not be dramatic; therefore, it is often ignored. However, the probability of a second concussion in the following period are greatly increased. Because the injuries are additive, the second impact may have dramatic consequences. If cerebral edema sets in, coma and even death will be the outcome.

Selenium: see trace elements.

Self pruning: the ability of the brain to downsize and then rebuild some of its complement of brain cells.

Serotonin: a neurotransmitter implicated in satiety and mood regulation (depression).

Shared Care: collaborative care between family physician and neuropsychiatrist.

Short term memory (STM): working memory is the shortest lasting seconds, as when remembering a telephone number. Working memory or STM is maintained by reverberating circuits in the cortex. Hippocampus encoded memory is longer and although it may last a few minutes, it is purged and forgotten. Any memory that is encoded through the hippocampus to form permanent traces is called long term or crystallized memory.

Silent infarcts: silent myocardial or cerebral infarcts.

Silent lobes: outdated and inaccurate term for the prefrontal cortex of the frontal lobes.

Sleep EEG: EEG tracing taken during sleep to study sleep architecture.

Sleep efficiency: the amount of time spent in real sleep divided by the amount of time spent in bed expressed as a percentage. For example, 9 hours spent in bed and sleeping 8 gives 8/9 x 100 = 89%.

Sleep hygiene: non-pharmacological practices that increase the likelihood of healthy refreshing sleep.

Sleep paralysis: the disconnection of the voluntary muscle system during dreaming (REM) sleep. Rarely a person may have a brief but usually frightening period of wakefulness yet feeling paralyzed. Also rarely, there may be no disconnection and bizarre acting out of dream material may ensue. That is known as REM associated behaviour disorder (RBD).

Sleep, sleep architecture: brain activity characterized by a succession of stages cycling through periods of deep slow wave sleep and lighter REM sleep. The patterns of these waves is referred to as sleep architecture. Normal sleep architecture is comprised of at least one long period of delta sleep and 4-5 shorter REM periods before awakening.

Slow wave sleep (SWS), delta sleep: deep sleep characterized by no dreaming and the activity of anabolic functions. The sleep EEG shows slow waves also called delta waves.

Snoring: noise made by the soft palate during sleep by partial obstruction of the airway.

Somatopause: a reduction of growth hormone levels at around and after menopausal years.

Somatotropin: see growth hormone.

Staying power of brain: see cognitive reserve.

Stearic acid: a colorless, waxlike, sparingly water-soluble, odorless solid. The most common fatty acid, occurring as the glyceride in tallow and other animal fats and in some animal oils. It is used chiefly in the manufacture of soaps, stearates, candles, cosmetics, and in medicine in suppositories and pill coatings.

Step counting: see Manpo-Kei.

Steroids: any of numerous naturally occurring or synthetic fat-soluble organic compounds having as a basis 17 carbon atoms arranged in four rings. They include sterols and bile acids, adrenal and sex hormones, certain natural drugs such as digitalis compounds, and the precursors of certain vitamins.

Stress busters: brief interventions or practices that lower cortisol levels.

Stress hormone: see cortisol.

Stress vulnerability: tendency to become stressed. Not able to avoid being stressed.

Stressor: anything that is causing a stress response.

Stroke: cerebrovascular accident (CVA), either hemorrhagic or ischemic.

Subarachnoid hemorrhage: bleeding into the space between the *dura mater* and arachnoid membrane brain coverings.

Subdural hematoma: bleeding into the subdural space, usually after head injury. The subdural space is between the membrane *dura mater* and the skull.

Substantia nigra: subcortical area rich in dopamine containing neurons dedicated to the modulation of involuntary muscle movements.

Synapse/neuron ratio: the number of synapses per brain cell expressed as a ratio. A related concept is synaptic density which refers to the number of synapses per cubic millimeter. Both numbers increase with learning activity and in healthy brain aging.

Synapses: the terminal buttons of axons and dendrites forming a pair of synaptic clefts where packets of neurotransmitter molecules are released affecting the transmission of information from one neuron to another.

Synaptic density: see synapse-neuron ratio. Related to cognitive reserve.

Synaptic plasticity: the formation, through gene expression and protein transcription, of new synapses in response to the stimulation of a neuronal pathway.

Synaptophysin: the brain derived neurotrophic (nerve growth) factor promoting new synapse formation.

Syndrome-X: see abdominal obesity.

Tartrazine: a yellow *azo* dye that is used in making organic pigments and in coloring foods and drugs. It sometimes causes bronchoconstriction in individuals with asthma. Eg., Yellow No. 5.

Testosterone: see androgen.

Theobromine: a white, crystalline, water-insoluble, poisonous powder, an isomer of theophylline and lower homologue of caffeine, occurring in tea and obtained from the cacao bean. It is used chiefly as a diuretic, myocardial stimulant, and vasodilator.

Thrifty gene theory: a theory that attempts to explain why mankind, and some tribes in particular, have a high propensity to weight gain. It upholds that we have an inherited, genetic tendency to store fat very efficiently because we have evolved this way, over thousands of years, to better able us to cope with severe food shortages during winter and frequent famines. Those who could store the most fat had a better chance of survival (James Neel, 1962).

Thrombosis: intravascular coagulation of the blood in any part of the circulatory system, as in the heart, arteries, veins, or capillaries or brain.

Thyroid gland: a gland in the neck which secretes thyroid hormone (thyroxin).

Thyroid hormone (thyroxin, thyroxine): the hormone secreted by the thyroid gland regulating metabolic rate.

Thyroid stimulating hormone (TSH): also known as thyrotropin, stimulates the thyroid gland to release thyroid hormone (thyroxin).

Thyrotropin releasing factor (TRF): a chemical messenger released from the hypothalamus to signal the pituitary gland to release thyrotropin (aka thyroid stimulating hormone (TSH)) which travels to the thyroid gland to trigger the secretion of thyroxin (thyroid hormone) into the bloodstream.

Tinnitus: ringing in the ears.

Tocopherols: any of a group of closely related, fat-soluble alcohols constituting vitamin E and similar compounds. Vitamin E family.

Total radical-trapping antioxidant potential in foods (TRAP values): a numeric value serving as a relative index of a foods antioxidant capacity.

Trace elements: elements required in very small quantities to enable the body chemistry to proceed faultlessly. See vital nutrients.

Transcutaneous electrical stimulation (TENS): electrical stimulation by electrodes on the skin to stimulate muscle contractions or sensory phenomena.

Transient ischemic attacks (TIAs): a temporary, brief loss of blood supply to a brain region producing symptoms specific to the function of that region.

Traumatic brain injury (TBI): injury to brain tissue as a result of an acceleration-deceleration injury, such as a whiplash or direct blow.

Treatment adherence: adopting and staying with a treatment plan. This is a newer term for 'compliance'.

Triage: the process of sorting victims, as of a battle or disaster, to determine medical priority in order to increase the overall number of survivors.

Underweight: low body mass index (BMI). See obesity.

Vascular dementia: neurocognitive impairment (dementia) caused by poor circulation and the cumulative effect of infarctions, mostly silent mini-strokes, in the brain.

Vascular diseases: diseases of blood vessels.

Vasculitis: inflammation of blood arterioles, arteries or veins.

Vasopressin: a hormone from the pituitary gland that plays a role in blood pressure regulation. It also has complex, poorly understood effects in some brain regions influencing learning, memory functions and bonding in mammals, notably, voles.

Vegetable oils: fats from plant sources. Most of these fats are unsaturated and in liquid form. Some, like olive oil, are not stable and break down if exposed to light and heat. Some, like palm oil, are saturated.

Vertigo: an unpleasant sensation of movement or spinning resulting from vestibular malfunction.

Vinclozolin: a fungicide.

Vital nutrients (minerals and vitamins): small amounts of these substances are required to enable the body chemistry to proceed in a healthy manner. Deficiency states cause characteristic syndromes.

Vitamin B$_{12}$ deficiency: characterized by anemia, memory problems, peripheral nerve lesions (numbness, tingling), poor balance and tinnitus.

Vitamin C deficiency: scurvy. Characterized by weakening of connective tissue leading to increased capillary fragility, easy bruising and bleeding.

Vitamin D: see calciferol.

Vitamins: see vital nutrients.

Waist circumference (WC): according to the National Institutes of Health, a high Waist Circumference (WC) is associated with an increased risk for type 2 diabetes, dyslipidemia, hypertension and cardiovascular disease when the BMI is between 25 and 34.9. Changes in WC over time can indicated an increase or decrease in abdominal fat. Increased abdominal fat is associated with an increased risk of heart disease. To determine your WC, locate the upper hip bone and place a measuring tape around the abdomen, ensuring that the tape measure is horizontal. The tape measure should be snug but should not cause compressions on the skin. For optimal health it should not be more than 40 in. (men) or 35 in. (women) – for Caucasians. Some interracial variations exist.

Waist to hip ratio: The waist-hip ratio is a more accurate predictor than simple waist circumference. Your waist to hip ratio is an important tool that helps you determine your overall health risk. People with more weight around their waist are at greater risk of lifestyle related diseases such as heart disease and diabetes than those with weight around their hips. It is a simple and useful measure of fat distribution. Use a measuring tape to check the waist and hip measurements. Measure your hip circumference at it's widest part. Measure your WC at the belly button or just above it. Risk for illness increases proportionately above 0.95 for men and 0.80 for women.

White matter of brain: brain areas deeper than the cortex characterized by nerve fibre (axon) bundles surrounded by white myelin, a cholesterol like substance that separates the axons and acts as an insulator to keep axons from touching each other.

Xanthines: a yellowish-white, crystalline purine base, that is a precursor of uric acid and is found in blood, urine, muscle tissue, and certain plants.

END

www.HealthyBrain.org